D0583537

ISBN 0947068030

Australian
REPTILES & FROGS

DISCARDED

Australian
REPTILES & FROGS

Raymond T. Hoser

50⁰⁰

OVERSIZE
QL
663
A1
H83
1989

Published by Pierson & Co
Distributed by Gary Allen

EL CAMINO COLLEGE
LIBRARY

Preceding page
Warty Green and Golden Bell Frog *Litoria raniformis*, Male,
Bundoora, Vic.

Published by Pierson & Co
P.O. Box 87
Mosman NSW 2088
Sydney Australia

In association with
Death Adder Services
2/1 The Close
Hunters Hill NSW 2110
Sydney Australia

Distributed by
Gary Allen Pty Ltd
9 Cooper Street
Smithfield NSW 2164
Tel (02) 725 2933

First published 1989

Copyright © 1989 Raymond Hoser

This book is copyright. Apart from any fair dealing as
permitted under the Copyright Act, no part may be
reproduced by any process without permission.
Enquiries should be addressed to the publishers.

National Library of Australia
Cataloguing-in-Publication Data
Hoser, Raymond T. (Raymond Terrence), 1962–
    Australian reptiles and frogs.
    Bibliography.
    Includes index.
    ISBN 0 947068 08 2.
    1. Reptiles – Australia. 2. Frogs – Australia.
    I. Title.
597.6'0994

Typeset by Bookset
Printed in Singapore by Singapore National Printers
Designed by Sandra Nobes
Edited by Cathryn Game
Photography by Raymond Hoser

CONTENTS

Acknowledgements   vii
Foreword   viii
Preface   ix

PART 1:   INTRODUCTION   1
Origins of Amphibians and Reptiles   1
Classification   3
Class Amphibia (Amphibians)   4
   Order Salientia (Frogs and Toads)   4
Class Reptilia (Reptiles)   7
   Order Crocodilia (Crocodilians)   8
   Order Testudines (Tortoises and Turtles)   8
   Order Squamata (Snakes and Lizards)   10

PART 2:   AUSTRALIAN FROGS AND REPTILES DESCRIBED   21
About the Descriptions   21
Class Amphibia   22
   Order Salientia (Frogs and Toads)   22
     Family Bufonidae (Toads)   22
     Family Myobatrachidae (Southern Frogs)   23
     Family Hylidae (Tree Frogs)   35
Class Reptilia   44
   Order Crocodilia   44
     Family Crocodylidae   44
   Order Testudines (Turtles and Tortoises)   46
     Suborder Cryptodira (Sea Turtles)   46
     Family Cheloniidae (Most Sea Turtles)   46
     Suborder Pleurodira (Sideneck Tortoises)   50
     Family Chelidae (Freshwater Tortoises)   50
   Order Squamata (Snakes and Lizards)   56
     Suborder Sauria (Lizards)   56
     Family Agamidae (Dragon Lizards)   56
     Family Gekkonidae (Geckoes)   66
     Family Pygopodidae (Legless Lizards)   79
     Family Scincidae (Skinks)   83
     Family Varanidae (Goannas/Monitor Lizards)   113

Order Squamata (Snakes and Lizards)   122
    Suborder Serpentes (Snakes)   122
    Family Typhopidae (Worm or Blind Snakes)   122
    Family Boidae (Pythons and Boas)   123
    Sub-family Pythoninae (Pythons)   123
    Family Acrochordidae (File Snakes)   140
    Family Colubridae (Colubrid Snakes)   140
    Family Elapidae (Front-fanged Venomous Land Snakes)   144
    Family Hydrophiidae (Sea Snakes)   175

PART 3:   CAPTIVITY AND CONSERVATION   177
Obtaining Reptiles and Frogs   177
Keeping Reptiles and Frogs   181
Feeding Reptiles and Frogs   184
Captive Breeding   189
Ailments, Diseases, Parasites and Their Treatment   194
Preserving Specimens   198
A Word of Warning   198
Photographing Reptiles and Frogs   199
Conservation of Reptiles and Frogs   202
    Reptile Habitats   206

References   231
Index   236

ACKNOWLEDGEMENTS

THE STUDY OF reptiles and amphibians is called herpetology. The name comes from the ancient Greek *herpeton* which meant 'crawling thing'. Nowadays the abbreviations 'herp' or 'herptile' are used when talking about reptiles and frogs, and people who study reptiles and amphibians are called herpetologists. 'Herpie' is a widely used term to describe anyone who has more than a casual association with reptiles and frogs.

Over the years many people have helped me in my herpetological endeavours, and either directly or indirectly helped me in the production of this book. My thanks to those named below, and my thanks and apologies to any who have been inadvertently omitted.

John Scanlon (NSW), Dr Richard Shine (NSW), Juanita Wright (Vic.), Cathryn Game (Vic.), and Katrina Hoser (NSW) read draft manuscripts and pointed out numerous errors, omissions, etc., as well as offering constructive comments. Any remaining faults in the text are my responsibility.

For the supply of reptiles and frogs to photograph, assistance in fighting corruption within the New South Wales fauna authorities, and other help, I would like to thank the following: Charles Acheson (NSW), Hugh Alexander (Qld), the proprietors, Australian Reptile Park (NSW), Val Bagshaw (WA), Richard Bartlett (USA), Joe Bredl (junior and senior) (SA), Dusty Brown (WA), William Bennett and family (NSW), David Carey and Glen Earnshaw (NSW), Michael Germac (Qld), Neil Charles (Qld), Harold Cogger (NSW), Ben Copeland (NSW), Robert Croft (NSW), Terrence Dawson (NSW), Alex Dudley (NSW), the Farmer family (NSW), Allen Greer (NSW), Gordon Grigg (NSW), Greg Hollace (NSW), Daryl Housten (NSW), Grant Husband (NSW), Bob Irwin (Qld), Doug Kirkner (Qld), John Lark (NSW), the proprietors, Manly Marineland (NSW), Scott Marx (NSW), Ted Mertens (SA), Peter Mirtschin (SA), the proprieters, Neptune's Coral Cave (Qld), William Miles (NSW), Greg Prostamus (Qld), Peter Richardson (Qld), Ron Sayers (USA), the proprietors, Royal Melbourne Zoo (Vic.), Richard Shearim (NSW), Ken Sheppherd (NSW), Gary Solazzo (Vic.), Gary Stephenson (NSW), Grant Turner (Vic.), Greg Wallis (NSW), Martin Wells (NSW), John Wilson (NSW), Steven Wright (NSW), Gary Zielfisch (Qld).

Robert Bredl (Qld), gave me access to his captive breeding records, many that were previously unreported.

Also I should thank the various journalists and lawyers who have assisted me in attempting to recover illegally stolen reptiles, files, photos and other miscellaneous items, often without success, though not through any lack of effort on their part. They include: Virginia Bell, Christopher Chapman, Fiona Cumming, Jeff Goldstein, Tony Gieuressivich, Michael King, Shane McGuire and others from 'Transmedia/Willisee', Tony Reeves, Michael Ramage, Barry Scott (all of NSW).

Ron Orders and Co. (UK) and the Northern Ohio Association of Herpetologists (NOAH) (USA) have assisted the cause at an international level. My brother Philip (a barrister) has often provided invaluable help. However, principal thanks must go to my parents Len and Katrina Hoser, who provided all assistance requested, and stood by me after I exposed major corruption in the New South Wales fauna authorities between 1973 and 1988. They also had to suffer the unforeseen consequences of my actions. In 1981 my parents became the first people to successfully recover snakes and files stolen by the New South Wales National Parks and Wildlife Service, giving much hope and encouragement to myself and herpetologists everywhere.

This book is dedicated to other herpetologists and the welfare of the reptiles and amphibians they seek to study.

FOREWORD

I HAVE BEEN closely involved with reptiles for over fifty years: for twenty years as a keen amateur in Europe, and for the past thirty years, in Australia on a professional basis. I am delighted to see the publication of Raymond T. Hoser's 'Australian Reptiles and Frogs'. Here is a book written in a manner that is easy to read, and which avoids overuse of the tedious and often confusing scientific classification. This book will encourage more people, both the young and the not so young, to take an interest in Australian reptiles and frogs. These animals play an important part in maintaining the delicate balance of Australian ecology, and for this reason I commend this book to you.

The author has travelled widely to collect material for the book, and he has photographed, in some cases for the first time, rarely seen specimens, by no means an easy task. The numerous excellent photographs will help the reader to identify the animals, but they will also, I believe, engender an appreciation of the beauty of Australia's reptiles. Readers will also find here many interesting and strange details about Australia's reptiles and frogs. Raymond Hoser's first-hand knowledge of most types of reptiles and frogs makes him well qualified to prepare this text.

One of the author's concerns in this book is to discuss the problems of conservation, endangered species, and keeping captive reptiles and frogs. The effects on reptiles and frogs of habitat destruction, feral pests and predators, can be devastating, and some species are endangered. Captive breeding may be the only means of preserving some species. Both the study and conservation of Australia's reptiles and frogs would be facilitated by the implementation of policy and legislation that would reserve protection for those species that *are* endangered. Without the stigma or fear of breaking the law by keeping these animals in captivity, often only presumed illegalities, many more people would take an interest in and undertake studies of our misunderstood reptiles and frogs. Captive breeding should be encouraged before it is too late to ensure the survival of any endangered species. Those who take an interest in reptiles and frogs will educate their neighbours, so that the notion that the only good reptile is a dead one, will become a thing of the past.

Legal and unobstructed distribution of reptiles and frogs bred in captivity would also significantly reduce the illegal trafficking in these animals, that has become such a problem and seriously concerns both wildlife departments and reptile keepers. The illegal export of wildlife must be stopped, and the only way to achieve this is for legal exportation of moderate numbers of specimens.

I recommend 'Australian Reptiles and Frogs' to all who are concerned with the future of our wonderful reptiles and amphibians, and to those who just have an interest in these creatures. I hope that this book will find many appreciative readers who will help save Australia's magnificent and beautiful reptiles and frogs for future generations to enjoy.

JOE BREDL
Bredl's Reptile Park and Zoo
Renmark, South Australia
Cardwell, Queensland
December 1988

PREFACE

AUSTRALIA HAS AN unfair surplus of reptile and frog numbers and species, yet we are behind most other parts of the world in relation to public knowledge about these animals, and research on them. We are in the dark ages in relation to conservation of these animals, and the information in this book aims to improve this situation.

In writing this book, I have endeavoured to provide information on a subject that has been somewhat neglected to date. This book deals mainly with all general aspects of Australian herpetofauna (reptiles and frogs). This book is not a strict taxonomic (classification) work, listing all species known. This has been attempted before. Those works, while useful, have suffered the problems of being outdated even before going to press, because new species are being continually described, and recognised species continually reclassified. Although dealing with the formalities of Australian reptile classification, I have avoided becoming bogged down on this subject and instead concentrated on providing a text that is easily readable by expert and interested novice alike.

I have dealt with a selection of most types of reptile and frog found in Australia. The 200-odd species dealt with here include those which are most commonly encountered, a cross section from all parts of the country, the best known species, and species of special interest for a variety of reasons.

I have attempted to keep the text as concise as possible, relying on photos to provide much stimulus to the reader.

In line with my views on conservation, I have provided photos of a number of habitats in which Australian reptiles and frogs live, which must be conserved in order to preserve the species within them. As Australia has the worst conservation record on a population basis of any developed nation, the principal aim of this book is to encourage more people to take an active interest in Australia's herpetofauna, before more of it disappears.

PART 1 INTRODUCTION

ORIGINS OF AMPHIBIANS AND REPTILES

AMPHIBIANS AND REPTILES, in the forms of frogs, tortoises and turtles, crocodiles, snakes and lizards, are recognisable to most people, yet most people know little of their origins.

Amphibians arose from the first vertebrates to leave the water and walk on land, about 400 million years ago in the Devonian period. Some labyrinthodont amphibians (which looked like giant salamanders) were as much as 3 metres long. The 'age of amphibians' lasted from about 380 million years ago to 270 million years ago. The first reptiles descended from labyrinthodont-like amphibians about 270 million years ago in the Carboniferous period. These four-legged carnivores gave rise to the dinosaurs during the Triassic, Jurassic and Cretaceous periods. The end of the Cretaceous period signalled the end of the 'age of reptiles'. After this period, mammals and birds, which both arose from reptiles, came to dominate the earth.

During the 'age of reptiles' enormous dinosaurs dominated the world's skies, seas and land. They included the swamp-dwelling Diplodocus measuring more than 25 metres and weighing 40 tonnes, the world's longest ever land-dwelling animal. The Brontosaurus, Supersaurus and others were the largest ever land-dwellers, weighing up to 60 tonnes. Tyrannosaurus was the largest land-dwelling carnivore known, being some 14 metres long and standing some 6 metres high. Giant bat-like Pterodactyls dominated the skies, while Pleiseosaurus and others took to the seas.

The period about 60 million years ago is one of about four periods of major extinctions known in geological

1 Death Adder *Acanthophis antarcticus*, male, red phase, head (West Head, NSW).

history, the others occurring earlier. The cause or causes of the relatively sudden decline of dinosaurs at the end of the Cretaceous period are not known, but might be one of or a combination of the following: warm-blooded more intelligent mammals killing and eating dinosaurs or their eggs; some kind of plant(s) poisoning and killing the herbivorous dinosaurs, leading to the starvation of the carnivorous dinosaurs which preyed on them; a disease or virus that killed off all the dinosaurs; a dramatic change in the world's climate, somehow disadvantaging the dinosaurs, although it is now believed that dinosaurs might have been endothermic (warm-blooded); meteorites hitting Earth and rapidly upsetting things; or some other unknown factor(s).

The first frogs appeared about 300 million years ago, but it is believed that these specimens still possessed tails. Tailless frogs of similar form to modern species were definitely present in the Jurassic period. Crocodiles and tortoises appeared in the Triassic period and have changed little since. Lizards appeared at the end of the Jurassic period some 140 million years ago, with the first known snake Lapparentophis appearing about 100 million years ago, towards the end of the age of reptiles.

Fossil snakes, lizards, tortoises and crocodiles, considerably larger than modern forms, are known.

The origins of many amphibians and reptiles are still unknown and current ideas are still largely conjecture.

Snakes presumably evolved from limbed ancestors, but the exact means by which this evolution occurred is not known. The possibility that all of today's extant snakes had more than one common ancestor, though unlikely, cannot be discounted. Some people have suggested that snakes evolved directly from land-dwelling reptiles, not unlike some 'Legless Lizards' now found. Others have suggested an aquatic origin, citing the long neck, lack of limbs and sometimes laterally compressed body. Still others have suggested a burrowing origin, citing structural evidence of snakes' eyes, skull characteristics, the disappearance of limbs and external ears. These people claim that the Asian Burrowing Monitor *Lanthanotus borneensis* is typical of some of the lizards that later evolved into snakes.

The theory of continental drift is now widely accepted. This theory states, among other things, that some 200 million years ago Africa, India, Australia, South America and Antarctica were all joined into a super continent called Gondwanaland, before drifting into their present positions. As a result of this, frog faunas of South America and Australia are similar in many respects, with both the Hylidae (Tree Frogs) and Leptodactylidae/Myobatrachidae (Southern Frogs) being the major frog types. Also, freshwater Tortoises of South America and Africa have a number of similarities to Australian species, all being 'side-necked' species.

The snake and lizard fauna of Australia appear to have evolved and diversified mainly since the break up of Gondwanaland, with most families being 'old world' rather than 'southern'. Two families, Pygopodidae (Legless Lizards) and Hydrophiidae (Sea Snakes), appear to have evolved in the Australasian or South-east Asian region in relatively recent geological times.

CLASSIFICATION

ALL LIVING THINGS are classified according to a single standardised system of classification, used by all biologists and other interested people. The system was developed by Linnaeus, the Swedish botanist Carl Linné (1707–78), and was first published in 1758 as a means of classifying living things in groups that would be understandable by all researchers.

The basic unit of classification is the species. The Darwinian definition of a species is 'a group of individuals which can freely interbreed and produce "normal" fertile offspring'. Typically, individuals of a given species are similar in appearance and habits. By strict definition, individuals of one species cannot interbreed with those of another species and produce viable young. But unfortunately in the real world, we often have problems defining a given species with these criteria alone.

Today a given species is defined using a number of criteria including geographical, chemical, physiological and genetic bases of reproductive isolation; and often there is still conflict among researchers as to which groups form species. As all groups of living things are continually evolving, there will always be situations where clines (intermediate forms) will be difficult to classify in an arbitrary, man-made category.

The modern classification system also tries to highlight evolutionary relationships between living things.

Similar species are placed in a genus; similar genera into a family; and so on to the kingdom level, of which there are about five recognised, including plants, animals, viruses, bacteria and blue-green algae, and protozoa. Within the classification levels are sometimes placed sub-species, species complexes, sub-genera, super genera and so on. These somewhat arbitrary divisions are used to clarify relationships between groups.

Names used for given species, genera, etc, are either Latin, ancient Greek or latinised. This is so that scientific names do not conflict with local colloquial names in any country. In theory, scientific names should remain constant, even when common/colloquial names for given living things change.

A given species is usually named according to a binomial system. The scientific name is always italicised. Two names are given. The first is the genus name, the second is the species name. Occasionally a third name is given, being the subspecies/race. The first letter of the genus name is a capital, while all other letters are lower case.

This is the system of classification used in this book. For a given species, the first described subspecies (the type) has the same as the species, while others are assigned new names. In some texts (including this), a person's name may appear in roman type after the species name. The name is of the person who originally described the species, and if the name appears in brackets, it means that the species described was originally placed in a different genus. The date after the name is the date which the species was originally scientifically described.

An example of the classification system at work is shown below, using the Scrub Python *Morelia amethistina*. However, it should be realised that the system is designed for convenience, and classification levels are usually only given when there might be doubt as to what one is actually discussing.

Kingdom: Animalia (animals)
Phylum: Chordata (chordates)
Sub-phylum: Vertebrata (vertebrates)
Class: Reptilia (reptiles)
Order: Squamata (snakes and lizards)
Sub-order: Serpentes (snakes)
Family: Boidae (boas and pythons)
Sub-family: Pythoninae (pythons)
Genus: *Morelia*
Species: *Morelia amethistina* (Schneider, 1801)
Sub-species: *Morelia amethistina amethistina* (found in New Guinea)
or:
Morelia amethistina kinghorni (found in Queensland)

When a given species is under continual discussion in a given written work, the scientific name may be abbreviated to initials for all but the last name given (e.g. *M.a. amethistina*).

Currently a lot of work is being done worldwide on the classification of reptiles, including Australian forms, and new species are still being described. The scientific names given in this book are those which are most widely agreed on by herpetologists at the time of writing. With a few exceptions, the scientific names used here correlate with those used by Cogger (1986). I have little doubt that as more work is done over the next few years some of the names will change.

CLASS AMPHIBIA (AMPHIBIANS)

AMPHIBIAN MEANS 'two lives'. Most amphibians are characterised by a distinctive aquatic larval staged called the tadpole. These tadpoles are typically similar in appearance and have large globular forebodies, with a tapering tail at the rear of the body. These tadpoles metamorphose into the adult form and then usually leave the water in which they developed. Typically tadpoles only develop in fresh water, and amphibians in general have a low tolerance of saline conditions.

Most larval amphibians possess gills, while adults are air-breathing and possess lungs. Most types of amphibian have a moist, 'slimy' skin, the obvious exception being dry warty toads, and the majority of species live in moist places. All are ectothermic (cold-blooded).

Three separate orders exist today. One is Gymnophiona, the caecilians, which are small worm- or snake-like burrowing animals found only in a few tropical areas outside Australia. The Caudata includes newts and salamanders which are found throughout the northern hemisphere. The only species likely to be seen in Australia is the Axolotl *Ambystoma mexicanum*, which is commonly kept as an aquarium pet, and only occurs naturally around Mexico City. The third order is Salientia (anurans). It includes frogs and toads, and is found throughout the world including most parts of Australia.

Order Salientia (Frogs and Toads)

Although about 2000 species are known, all frogs and toads are of similar form (as will be seen in the illustrations).

With the exception of toads (Bufonidae), frogs usually have a moist, highly permeable skin and, therefore, in order to avoid dehydrating must live in moist places. Even desert species are usually found close to water, and are usually only active after rain. Most anurans are also nocturnal, probably in a bid to minimise exposure to the drying rays of the sun, and exposure to diurnal predators such as birds. A few frog species from colder places are known to actually bask in the sun, when concealed, adjacent to watercourses (e.g. *Litoria aurea*).

The moist skin of frogs assists in thermo-regulation and gas exchange. The surface layers are shed periodically. The chromatophores (colour cells) are able to change colour in response to various factors, and often help frogs to change colour in response to the environment, and therefore avoid predation. Poisonous frogs are often brightly coloured, to warn predators of their nature. Poison is secreted by some frogs from glands situated on various parts of the body. The main source of poison on the Cane Toad *Bufo marinus* are the parotid glands which are situated conspicuously behind the eyes on the back of the head. Frogs' senses of sight, hearing and touch are usually well developed, with most species being able to 'croak' or call one another for breeding purposes or when attacked.

2   Tadpoles (larval frogs), *Limnodynastes peronii* (Lane Cove, NSW).

Frogs have very keen eyesight which is important for them in catching their food. As they are generally nocturnal, they tend to have eliptical eye pupils.

Most species are indiscriminate in what they eat, feeding on all types of terrestrial arthropods small enough to fit in their mouths. Larger species will feed on other frogs, small reptiles and sometimes even small mammals. A few species of frog appear to feed exclusively on termites, but their biologies are little known.

Frogs' food is captured by the sticky muscular tongue which is mounted at the front of the mouth and loose at the rear. The forelimbs are used to help stuff larger food items into the mouth.

3   Metamorphasing tadpoles, *Limnodynastes peronii* (Lane Cove, NSW).

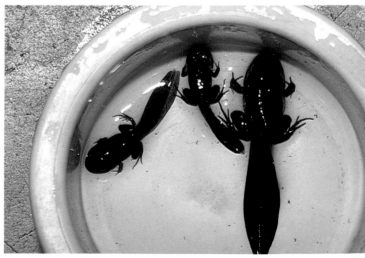

Frogs are eaten by a number of other vertebrates, particularly snakes and birds. Tadpoles are preyed on by fish and other aquatic animals, and bodies of water with large numbers of fish usually lack tadpoles.

Most species of frog take a year or two to reach maturity and they can live for several years. Some larger species have been known to live for up to twenty years in a predator-free captive environment.

LOCOMOTION

Frogs are found throughout most parts of the world in large numbers and their success is no doubt due in part to their jumping locomotion, executed by their elongated hind limbs.

The frog species credited with being the longest jumper is an African species of the family Ranidae, *Ptychadena oxyrhynchus*, which is known to be able to jump 10 metres in three successive leaps.

Not all frogs hop, however. Some genera, including *Pseudophryne* and *Myobatrachus*, more often walk on all fours, rather than hop.

Some frogs (e.g. *Cyclorana* spp.) are able to burrow by using their hind limbs as shovels, and moving rear-first into soft mud. They can then entomb themselves in moist chambers and survive drought periods. These frogs are called waterholding or burrowing frogs, are typically stout in build and usually live in drier environments.

Tree frogs (Hylidae) have developed 'suction caps' on their toes which enable them to climb all types of surface including glass, thereby opening up a range of new habitats for themselves. The success of this adaptation is evidenced by the large number of species of tree frogs found throughout the world.

4 Giant Burrowing Frogs *Helioporus australiacus*, adult male and juvenile (West Head, NSW).

5 Green Tree Frog *Litoria caerulea*, head (Wyong, NSW).

6 Northern Banjo Frog *Limnodynastes terraereginae*, head (Yamba, NSW).

7   Green Tree Frog *Litoria caerulea* (Charters Towers, QLD).

BREEDING

Frogs are efficient swimmers and usually return to water to breed, where they mate and lay eggs. In warmer places frogs breed during wet seasons, while in colder climates they usually breed in spring and summer, when it is warm enough to be active, and waters are not frozen.

To attract a mate, male frogs will usually call using well-developed vocal sacs at the bottom of the mouth. This vocal sac can be enlarged considerably when the frog is calling.

Fertilisation is external and the mating position is called amplexus. In amplexus the male frog positions himself on top of the female, by holding on to her with his forelimbs and fertilising the eggs with his sperm as they are laid. The eggs usually hatch about two days later, and tadpoles typically take between six weeks and three months to grow and metamorphose into small frogs.

Some frogs (e.g. *Limnodynastes peronii*) beat their egg mass into a foamy floating mass when laying them. Others lay their eggs in characteristic strings attached to submerged vegetation (*Litoria phyllochroa*), while others (e.g. *Mixophyes* spp.) lay their eggs on land adjacent to water and when they hatch the tadpoles make their way into the water.

Some frogs have other unusual breeding biologies. One species (*Rheobatrachus silus*) has its tadpoles develop within the stomach of the adult frog. Other species (*Kyarranus* spp., *Philoria frostii*) have their eggs develop in a foamy mucous mass in a burrow away from water. Their tadpole stage is repressed and hatching coincides with metamorphosis.

Tadpoles are usually omnivorous, feeding on small detritus, algaes, etc. Some species, however, are strongly carnivorous and eat other tadpoles, or even tadpoles of their own species. Cannibalistic tadpoles often become cannibalistic frogs.

MAIN DIAGNOSTIC FEATURES

To classify frogs the following diagnostic features are commonly used:

1 Morphology, including size, dentition, feet structure, etc.

2 Colour(s).

3 Tadpole morphology, etc. (particularly colour and mouthparts construction).

4 Call. All calling species have a unique call; very few types are not known to call.

AUSTRALIAN FROG FAMILIES

In Australia there are currently five recognised families of frog. These are:

1 Hylidae — Tree Frogs (found Australia wide).

2 Myobatrachidae (formely included in the family Leptodactylidae) — Southern Frogs (found Australia wide).

3 Ranidae — True Frogs (found north Queensland).

4 Microhylidae — Narrow-mouthed Frogs (found north Queensland).

5 Bufonidae — Toads (introduced to Queensland; now found Queensland, Northern Territory and northern New South Wales).

CLASS REPTILIA (REPTILES)

REPTILES HAVE A number of distinctive features. All possess a scaly skin, which is shed periodically, to facilitate growth. Under 'normal' circumstances all modern reptiles are ectothermic (cold-blooded), meaning that their body temperature is regulated by that of their immediate environmental surroundings and by direct solar radiation.

Unlike amphibians, reptiles have no larval stage. Most young reptiles are miniature replicas of their parents and all are capable of fending for themselves at birth. With the exception of crocodilians and pythons, parent reptiles do not care for the young.

Although about a hundred orders existed during the age of dinosaurs, only four orders exist today. They are:

1 Rhynchocephalia — represented by a sole species, the lizard-like Tuatara of New Zealand.

2 Crocodilia — crocodiles, alligators and gavial.

3 Testudines — tortoises and turtles.

4 Squamata — snakes and lizards.

All four groups were present during the age of dinosaurs and are therefore ancient, although the squamates have undergone most of their adaptive radiation (speciation) during the last 60 million years and can therefore be termed truly modern reptiles.

Most reptiles have some courtship behaviour before mating, with each order of reptiles having relatively stereotyped sexual behaviour (discussed later). In most reptiles, the males are the ones who appear to initiate mating.

The rhynchocephalia apparently have no copulatory organ. Testudines and crocodilians have a penis, whilst squamates have paired copulatory organs called hemipenes, only one of which is used at a time depending on the angle from which copulation takes place. The hemipenes are analogous to the penis in that they evolved independently to serve the same function. Hemipenes normally are held within the base of the tail, and they evert outwards through the anus (vent) when copulation takes place. Hemipene structure varies with each species, and is sometimes an important classification feature. Some species have bi-lobed hemipenes, while many others, usually snakes, have 'spikey' hemipenes. These spikes help hold the single hemipenis within the female during copulation.

Another feature which separates reptiles from most mammals is the fact that under normal circumstances the genitals are held within the body cavity.

In all reptiles fertilisation is internal. All reptiles have some capacity for sperm storage, although this ability varies with different types, and this area needs further research to clarify the exact sperm storage ability of female reptiles.

With the exception of some snakes and some lizards, all reptiles lay eggs, which require incubation within particular temperature and humidity constraints to produce fully-formed young. All hatching reptiles possess an egg tooth on the tip of the snout which is shed shortly after hatching. In the case of some types of crocodilian, lizard and testudine, incubation temperature of eggs dictates the sex of hatchlings instead of genetic determinants. In these species above averge incubation temperatures (at specific stages of incubation) usually result in females, while below average temperatures usually produce males.

The cold-blooded nature of reptiles should not be regarded as an anachronism. It is an adaptation to minimise metabolic energy used during inactive periods, and the slower metabolic rate of reptiles often enables them to survive in areas where warm-blooded animals cannot.

When active, the body temperature of some reptiles may approach 40°C, while some inactive reptiles are known to survive below freezing point. Reptiles can maintain desired body temperature by regulating where they position themselves within their environment. In most areas reptiles will bask in the sun or on warm ground, in order to raise their body temperature, and hide under cover if it gets too hot. Reptiles may flatten out their bodies on the ground surface when basking in order to maximise exposure. When overheated reptiles may pant, by holding their mouth open and allowing air to circulate around their inflated body, in a bid to aid cooling.

Some reptiles are able to adjust their colour for a number of reasons. Generally they become darker when cooler, and lighter when warmer. Some reptiles, such as many Agamid lizards, can change their colour almost instantly in response to temperature, while others, such as the Inland Taipan *Oxyuranus microlepidotus*, change colour on a seasonal basis.

When climatic conditions become excessive — too cold, too hot or too dry — reptiles may go into aestivation or hibernation. This is essentially a state of inactivity or torpor, whereby the reptile places itself in the most suitable site to await the return of more favourable conditions. This is usually under deep cover or underground where adverse temperature fluctuations are at a minimum. The reptile will emerge with the return of favourable weather. Hibernation is usually a seasonal occurrence in areas where reptile species are known to do it.

The growth rate of reptiles varies strongly both between species and between individuals of a given species. Growth is largely dependent on food intake for wild specimens, with specimens that obtain more food growing faster than those that obtain less food. In captivity, however, growth rates can be further accelerated by increasing temperature at which the reptile is kept, as well as by boosting food intake, as this serves to put the metabolic rate at a maximum. It is possible to double wild growth rates in captive specimens of many reptiles.

Although it is often stated that reptiles continue to grow throughout life, studies show that older specimens do, in fact, stop growing. A particularly large reptile of a given species is not necessarily an old reptile. Reptiles' sizes vary as a result of genetic factors in the same way that sizes of adult humans and other animals vary.

The presence of a cloaca which enables reptiles to resorb most moisture and pass nearly solid faeces, and their lack of sweat glands, enables reptiles to colonise arid areas successfully. In these areas, snakes and lizards are often the dominant vertebrates.

Reptiles might be diurnal, crepuscular (dusk active), nocturnal, or two or three of these. Generally speaking, in most parts of Australia most lizards are diurnal, while most snakes are nocturnal.

Usually diurnal species have round eye pupils, while nocturnal species have elliptical pupils.

Order Crocodilia (Crocodilians)

All modern-day crocodilians belong to the family Crocodylidae and I have dealt with many of their general characteristics in my description of the family later in the book.

Crocodilians are distinguishable by a number of characteristics. The four-chambered heart is unique among the reptiles (but not unlike mammals), and is regarded as being superior to the three-chambered heart found in other reptiles as it separates oxygen-rich blood from oxygen-deficient blood.

The nasal passage of crocodilians extends to the back of the throat. There is an effective valve at the base of the tongue which enables crocodilians to swallow prey below the water surface and still breathe through its nostrils above the water surface.

Crocodilians have the ability to remain submerged with only their eyes and nostrils protruding above the water surface. They are capable of producing a loud hiss and a growling sound, and young specimens often let out a high-pitched yelp.

The stomachs of crocodilians always seem to contain some loose stones which apparently serve to aid in digestion and to act as ballast for buoyant bodies.

Crocodiles' eggs usually take about two or three months to hatch, and young take several years to mature. Life spans of more than forty years are known.

MAJOR DIAGNOSTIC FEATURES

With only about twenty-three species found throughout the world, people rarely have difficulty identifying given species. Distinguishing features include:

1 Location found.

2 Morphology, particularly that of the snout and the dentition.

3 Head scalation.

Order Testudines (Tortoises and Turtles)

Characterised by their usually hard horny shells, which enclose the body, these relatively slow-moving reptiles are found in most parts of the world including most parts of Australia. Of about 300 species found, some thirty or more species occur in Australia and adjacent seas.

Although aquatic, all species may drown if held underwater for too long, although some freshwater species apparently hibernate, submerged in mud at the bottom of rivers, during colder periods.

Tortoises and turtles are among the longest lived of reptiles. Reliable records for some specimens show life spans in captivity exceeding a hundred years. Some Australian species are known to live for forty years or more.

The dorsal surface of the shell is called the carapace, the ventral surface is called the plastron, and the adjoining parts are called the bridge. The horny plates of the shell are the equivalent of the scales found in other reptiles.

Most species are omnivorous, but the majority tend to eat more meat than vegetable material. All modern species lack proper teeth, having their jaws lined with sharp ridges which form a horny beak. Larger specimens can give a very painful bite and some have been known even to sever a finger.

DEFENCE AND PREDATION

The shell is the primary protection for most species when adult, and all have some degree of ability to withdraw into the shell, although in sea turtles this is minimal.

The vulnerable parts of a sea turtle are the flippers, so when attacked by sharks and other predators in the sea, sea turtles will float on the sea surface with their flippers raised above the water level, in a bid to protect their flippers. Sea turtles with parts of their flippers missing are common. Marine turtles are virtually helpless on land and consequently only come ashore to lay eggs, usually at night in order to avoid predators. When young specimens hatch, all leave the nest at the same time, presumably in a bid to minimise the mortality rate.

Although both sea turtles and larger freshwater species have strong jaws, they are only used as defence against humans, when they are handled.

Freshwater species possess odour glands adjacent to the bridge of the shell. Some species, including the Long-necked Tortoise *Chelodina longicollis*, release an unpleasant odour when attacked. Captive specimens soon stop doing this when they become used to being handled and their new surroundings.

Both freshwater and marine species are often caught and killed by fishermen, in nets, and by those who kill them in the belief that they eat fish. It should be noted that all types would eat very few fish in relation to the number of fish present at any given time, and usually only the slow and sick fish, as most testudines are too slow-moving to catch them.

Small and young tortoises and turtles are killed and eaten by a variety of mammals and large lizards, which may also dig up nests and eat eggs and/or hatchlings.

BREEDING

Both marine and freshwater testudines are capable of producing several clutches of eggs from a single mating, even over several years. It is assumed that they have the ability to store sperm. In many species of marine and short-necked freshwater species, males are recognisable by their larger tails, and in some marine and freshwater species by their concave plastrons.

Courtship consists of the male following the female wherever she goes, and sometimes circling her, attempting to mount her, etc. This may last for up to two months.

Mating is usually done in shallow waters (except in non-Australian land tortoises) and many individuals are caught and/or killed when preoccupied with mating. Copulation occurs when the male mounts the female and lowers the rear end of his shell, so that the two tails meet. The male's penis is then inserted into the female's tail. The copulation act can last up to an hour or more.

There are variations in the nesting patterns of marine turtles and freshwater tortoises.

Most adult marine turtles only nest every few years, but produce multiple clutches of eggs in the years that they breed. For example, the Green Turtle *Chelonia mydas* may lay up to five clutches of eggs in a season, with about 110 eggs per clutch.

Turtles appear to nest on the beach on which they were born, and often travel thousands of kilometres to do so from their feeding grounds elsewhere. How they find their way back to a particular beach is not known.

During the nesting season, a good nesting beach (rookery) is easily recognisable by the wide tracks left in the sand by the nesting turtles.

Often turtles hover in the water around their nesting beach in small schools, before nesting at the high tide at night. Turtles will often come ashore several times (on one or many nights) before actually constructing a nest and laying eggs. The eggs are typically laid immediately behind the high tide mark.

On finding a suitable nesting site, the female turtle creates a saucer-like depression by using her flippers to rotate the body into the sand. The depression is slightly larger than the turtle in diameter (the depression being usually over a metre in size). The hind feet are then used to dig the nesting hole at one end of the depression, and by curling the feet she slowly lifts sand and throws it out of the nest one footful at a time. This tedious process can take up to two hours until the nest is completely dug.

Once the hole is dug as deep as the turtle's rear flippers will go, the turtle drops her ping-pong-ball-like eggs into the pear-shaped hole. It generally takes about ten to twenty minutes to lay all her eggs. On completion of laying the turtle fills in the hole and pushes more sand over the nest until a low mound results. By shifting her body and throwing sand with her flippers the turtle attempts to conceal the whereabouts of her nest. She then returns to the sea.

Throughout the procedure, from emergence from the sea until its return, the turtle rests frequently. The whole procedure obviously requires a huge amount of effort and is painfully slow.

Depending on sand temperature and other factors, eggs usually take about six to eight weeks to hatch. The young turtles, when hatching, wait until all in the nest have hatched before deciding to leave the nest. Then, usually at night, when the sand temperature is relatively cool, the turtles will make for the sea *en masse*. The time between the first turtles emerging from the nest and the last reaching the sea is usually only a few minutes. The small turtles know to head for the sea by the increased light present over the sea, as opposed to land, due to the reflections over the water. However, artificial or human lights can create havoc for hatchling turtles. Young turtles will congregate around man-made lights or follow a torchlight whichever way it goes instead of making their way to the sea.

Freshwater tortoises have been reported by Aboriginals as nesting on high sandbanks of rivers, but it is not known whether freshwater tortoises attempt to return to their birthplace to nest, although it is believed to be unlikely. Most freshwater species appear to nest in suitable sites near where they happen to be at the time. These sites are usually located above the flood level. Female tortoises will nest up to 500 metres from water, or climb steep banks more than 100 metres high in order to find suitable sites. Egg clutches, which might number from four to twenty-four, are laid in a similar manner to that of sea turtles, although freshwater tortoises are more mobile, and will dig proportionately deeper holes in which to deposit their eggs.

Some long-necked species, including the Broad-shelled Tortoise *Chelondina expansa*, commonly make a hard earthen plug to seal the nest. This is done by passing fluid from the cloaca into the surface dirt of the nest so that it hardens like mud.

Freshwater tortoises are experts in concealing nests. I had a number breeding in a large pit for several years, and often the only knowledge I had of breedings was when I saw hatchlings in the pond. More often than not I was unable to find the nests from which these tortoises emerged, and this was only in a 6-metre-by-6-metre pit.

When captive specimens have deposited eggs in water, it is usually because they have been unable to find what is deemed by the tortoises to be a suitable nesting site.

Most freshwater tortoises only emerge from nests after rain and, if a season is particularly dry, hatchlings may die within the nest. Young Broad-shelled Tortoises have been known to remain within a nest during a drought for more than twelve months after hatching.

Incubation periods for eggs range from three to twelve months depending on the species and incubation temperature.

8   Krefft's Tortoise *Emydura krefftii* (Nambour, QLD).

DIET AND FEEDING

All Australian testudines (marine and freshwater) appear to be mainly carnivorous, feeding on small fish, crustaceans and various other soft-bodied animals. In the wild the majority of species probably obtain most of their food by scavenging for dead remains, etc. All species also appear to feed on some plant material, such as water or sea weeds, and some species are known even to feed on the nut from Pandanus palms in northern Australia.

In captivity all species feed readily, and once they are settled in will readily attempt to feed on the fingers of their owner if given the opportunity. How obese a testudine can become is restricted by its shell and, other than by weighing a testudine, it is hard to gauge how well fed it is.

AUSTRALIAN TESTUDINE FAMILIES

There are currently four families of testudines found in Australia. They are:

1 Cheloniidae — most Sea Turtles (found in northern Australian seas).

2 Dermochelyidae — Leathery Sea Turtle (found in northern Australian seas).

3 Carettochelydidae — Pitted Shell Turtle (found in rivers in the Northern Territory).

4 Chelidae — Freshwater Tortoises (found in most of mainland Australia).

MAIN DIAGNOSTIC FEATURES

The main diagnostic features used to classify testudines are:

1 Morphology, particularly that of the shell.

2 Scalation of the head.

Order Squamata (Snakes and Lizards)

Snakes and lizards are by far the most abundant reptiles in Australia and the world today. About 5000 species are known worldwide with about 500 found in Australia. Although no snakes have legs, some lizards also lack legs or have greatly reduced limbs, and therefore one has to find characteristics that separate snakes from legless lizards. These characteristics are as follows:

1 Snakes have wide belly (ventral) scales, while lizards have small belly scales, not much larger than those dorsally. (Blind Snakes [family: Typhlopidae] are an exception to this, but they are very distinctive in appearance.)

2 All snakes have a forked tongue, while only a few types of lizard with well-developed legs have forked tongues.

3 Snakes have tails much shorter in length than their body, whereas legless lizards usually have much longer tails than their body (unless the tail has been shed, which would be obvious anyway).

4 Snakes lack an external ear; most lizards have a visible external ear.

5 Snakes lack eyelids and the eye is covered by a fixed transparent scale. Most but not all lizards have movable eyelids.

6 With the exception of Blind Snakes (Typhlopidae), snakes' jaws are loosely attached with elastic ligaments, whereas lizards jaws are rigidly sutured.

Both snakes and lizards possess a sensory device called the Jacobsen's organ, which aids the sense of smell. This paired structure consists of two odour-sensitive pits located at the front of the roof of the mouth. Scent particles are transported to these pits from the air or elsewhere by way of the tip(s) of the tongue. Forked-tongued lizards and all snakes tend to flicker their tongues constantly when active, due to the importance of the Jacobsen's organ for the location of food, defence, etc.

Little research has been done on the longevity of squamates, although it is believed that most larger species can live for up to twenty years.

Because of the relative inelasticity of the scaly skin, and the need for growth and general body maintenance, the entire outer skin is shed at periodic intervals. In most lizards this is usually done in a piecemeal fashion, while snakes usually shed the skin in one piece. Snakes and legless lizards do this by simply moving out of the skin and leaving the inside-out epidermal cast behind. In snakes, the transparent eye scale is shed with the rest of the skin. This scale becomes opaque before sloughing, leaving the snake effectively blind. The eye scale usually clears about 24–48 hours before actually shedding. Before sloughing, a film of moisture develops between the new skin and the old (which also causes eye clouding in snakes). The moisture film serves to ease sloughing for squamates and they have trouble doing so when it does not develop.

Rapidly growing reptiles shed their skins more often than those which are not growing. Reptiles afflicted by skin ailments, including parasites, will also shed their skin more frequently. Rapidly growing snakes have been known to slough at one-month intervals over a twelve-month period. Most adult snakes slough every three to four months under normal circumstances. No studies have been done on the frequency of sloughing in lizards.

Both snakes and lizards occur in a wide range of colours, including greens, blues, yellows, reds, blacks and combinations of these. These colours may result from camouflage, as a warning to would-be predators, thermoregulation needs, or other reasons.

Reproductive modes of snakes and lizards range from oviparous (egg-laying), ovoviviparous (eggs developing within the body, and live young being produced), viviparous (live young developing within the body of the female which derive nourishment during development from the female: similar to the mammal's placental system), and various intermediate levels.

Even within a single genus, reproductive modes vary. The live-bearing mode has evolved on a number of occasions in both snakes and lizards. Live-bearing reptiles typically produce their young when weather conditions are most favourable for the survival of the young.

Egg-layers typically lay their eggs in sheltered, warm and moist places. Some types of snake and lizard lay eggs in communal sites with up to 200 eggs being laid in a given site, by many individuals. Usually, but not always, only one reptile species will utilise a given communal nesting site. Eggs laid by reptiles range from those with thin, parchment-like membranes for a shell, to those with thick calcified shells, depending on the species. Unlike crocodilians and testudines, squamate hatchlings do not appear to wait for all eggs to hatch before leaving a nest.

Snakes and lizards are known to produce a given batch of young or eggs over periods ranging from a few seconds to several days. Incubation times for squamate eggs range upwards from almost immediately in a few species to five weeks or even nine months, depending on the species. Most species' eggs tend to hatch some six to twelve weeks after being laid. Those species which have eggs that hatch shortly after being laid produce eggs with almost fully-formed young.

Smaller species of squamate usually mature within two years, while larger species usually take slightly longer. Most types can live for more than ten years, with some species having recorded life spans of up to forty years in captivity.

Suborder Sauria (Lizards)

In Australia there are five lizard families:

1 Agamidae — Dragons (found Australia wide).

2 Gekkonidae — Geckoes (found in all mainland Australia).

3 Pygopodidae — Flap-footed Legless Lizards (found in all mainland Australia).

4 Scincidae — Skinks (found Australia wide).

5 Varanidae — Monitors/Goannas (found in all mainland Australia).

Two species of lizard in North America are venomous (*Heloderma* spp.), but no Australian lizard is venomous.

TAIL AUTOTOMY

A unique feature of many lizards is tail autotomy. This is the ability of lizards to loose their tails when under threat. Pressure applied to the tail by a predator allows the tail to break off, usually at the point where the pressure is applied. The tail then wriggles vigorously distracting the predator while the lizard escapes. The lizard is apparently unharmed and it regenerates a new tail in time. The new tail or tail portion usually lacks the length and colour of the original tail, and might sometimes be of remarkably different appearance.

The survival value of this adaptation is apparently huge, judging by the number of skinks and other lizards caught with regenerated tails.

Sometimes the tail does not completely break off when attacked at a given point. This can sometimes result in the original tail surviving and a new tail growing from the point of the break. On other occasions more than one tail can regenerate from a clean break. Some lizards have been known to possess up to seven tails.

In Australia all pygopids, geckoes, most skinks and a few agamids are capable of tail autotomy.

9   Copper-tailed Skink *Ctenotus taeniolatus* with two tails (Pearl Beach, NSW).

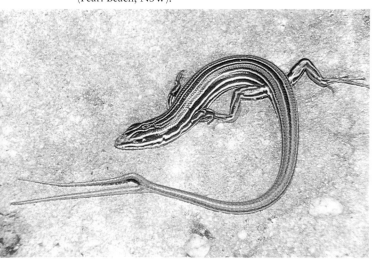

DEFENCE

Besides autotomy, lizards employ a number of other means to avoid predation. Most species are very fast-moving and are simply very hard to catch if they choose to flee.

Other species, particularly slower types, have evolved a number of convincing threat displays. Blue-tongue lizards (*Tiliqua* spp.) when cornered will stand their ground, mouth agape, and reveal a brightly coloured throat and tongue. They will then lunge at the aggressor holding the mouth wide open, with the tongue placed at the front of the mouth. Simultaneously the lizard will inflate and flatten out the body and hiss menacingly. Frill-necked Dragons *Chlamydosaurus kingii* and Bearded Dragons *Pogona* spp. will employ a similar strategy, erecting their frill or beard in a bid to make themselves appear larger than they really are.

Some species, such as the Forest Dragon *Gonocephalus spinipes*, rely entirely on cryptic colouring to avoid detection, and when approached usually remain completely still. Earless Dragons *Tympanocryptis* spp. flatten out their bodies at ground level when approached.

Most lizards when pursued make for cover which is inaccessible to would-be predators. This is commonly rock outcrops and boulders, where they hide in crevices, while other species will take to trees or even water when threatened.

A number of skinks (Genus *Egernia*) have developed spines on their bodies. They are able to move into crevices and puff up their bodies and, because of their spines, become very difficult to dislodge.

Some legless lizards (Pygopodidae; *Delma* spp.) will, if startled when foraging, adopt an unusual means of escape. Using very rapid movements they will stand up on their tails and jump away at high speed, before resuming their normal mode of locomotion.

Some legless lizards occur in similar colours and markings not unlike those of venomous snakes (e.g. *Pygopus nigriceps*, *Delma tincta*), in a bid to confuse and deter potential predators.

When caught many lizards will defecate. This also helps to repell predators.

FEEDING

Most skinks, dragons, geckoes and some pygopids actively stalk and capture living prey, usually in the form of arthropods and other small animals. Some larger types, including Bluetongues *Tiliqua* spp. and larger agamids, are omnivorous, feeding on vegetable matter and various forms of meat as well as the usual arthropods, etc.

Most large monitors feed on a range of vertebrates and carrion only, while smaller monitors and a few pygopids feed almost exclusively on smaller reptiles, which they actively chase or ambush.

Most lizards, particularly the geckoes, use the tail as a food store during lean periods. The tail sometimes appears very bulbous.

Dragons (Agamidae) possess sharp pointed teeth, which are generally useful for chopping up their insect prey, and various plant materials. Monitors (Varanidae) have thin, sharp and backwardly-curved teeth which are useful in holding on to moving prey and when used in conjunction with their claws are useful in tearing it apart. Skinks', geckoes' and pygopids' teeth are covered with

10 Blotched Bluetongue *Tiliqua nigrolutea*, NSW alpine form, head (Lithgow, NSW).

sharp horny ridges, which serve to crush their prey when necessary. Larger skinks with powerful jaws are capable of giving very painful bites.

FIGHTING

Male lizards will often compete for a given territory or a female.

Some lizards, including the Lace Monitor *Varanus varius*, have male combat during the breeding season. Males will fight, often inflicting wounds on one another, before mating with females. The male that appears to lose the fight often still copulates with the female being fought over. In these fights goannas use all physical means to hurt one another, including clawing and biting.

Some lizards, notably some agamids, will engage in male-to-male rivalry using threat displays, as opposed to actually fighting. Some dragons employ other unusual

11 Jacky Lizard *Amphibolurus muricatus*, male, cooled specimen, darker in colour (same specimen as that pictured in Fig. 129).

behaviour when engaging in male-to-male rivalry. This includes head bobbing, tail lashing (from side to side), and arm waving (usually the forearms). In these lizards often similar behaviour is used to attract the female before copulation.

Male lizards of other species may use a combination of tooth and claw fighting and threat displays when rivalling for a mate or a territory. Male Bearded Dragons *Pogona barbatus* will circle one another using standard threat display (mouth agape) intermittently. They will also attempt to bite one another's tails, jerking the other lizard at the same time. The loser is the lizard who flees at the end of the 'game'.

MATING

In some types of lizard, males and females are of different colours. Usually the males are brighter in colour, often with bright flushes of red or blue, usually on the sides and belly. These bright colours are obviously an aid to attracting mates, and aid in sexual identification for the lizards themselves.

Some lizards have apparently simple mating behaviour, while others have elaborate courtship patterns before copulating.

The actual mating for all lizards appears to be the same. The male simply mounts the female then positions his vent adjacent to the female's, by twisting the rear part of his body so that his hind parts are underneath those of the female. By twisting his tail around that of the female, and using his legs, he is able to maintain his position in relation to the female lizard. The male then everts one hemipenis and then, usually after a bit of difficulty, it is placed within the female, where it swells, and the two lizards then become connected. To separate two copulating lizards can be very painful for both of them. After the male completes the sperm transfer, the hemipenis shrinks and is then withdrawn back into his body.

REPRODUCTION

A few types of lizard, including most specimens of the Bynoe's Gecko *Heteronotia binoei*, are parthenogenetic. That is, all specimens are females and mating is not necessary. This trait appears to have evolved as a result of genetic mutation. Individuals of parthenogenetic species appear to have large numbers of chromosome sets, far more than the usual '2N' pattern (3N upwards).

Egg-layers typically lay eggs in especially dug nests, in warm moist places or under cover, or both. Like all reptiles, lizards usually take considerable care in concealing their nests. The number of eggs laid ranges from one to thirty, depending on the species. Live-bearing lizards typically produce less than ten young, although a few species may produce up to twenty or more young.

MAIN DIAGNOSTIC FEATURES

Due to the number of species, and the variety of lizards, a number of diagnostic features are used to identify given species. They include:

1 Morphology.

2 Colour.

3 Size.

4 Scalation (particularly the head).

5 Sexual characteristics, including pre-anal and femoral pores (on hind limbs), near tail swellings, etc.

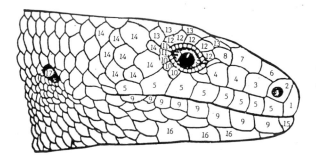

12 Mertens' Water Monitor *Varanus mertensi* (locality unknown).

Scalation of a skink lizard's head (lateral view).
(Pink Tongued Skink *Tiliqua gerrardi*).
1. Rostral 2. Nasal 3. Post nasal 4. Loreal
5. Supralabial 6. Frontonasal 7. Prefrontal 8. Canthal
9. Infralabials 10. Postsubocular 11. Postocular
12. Supraciliary 13. Supraocular 14. Temporal
15. Mental 16. Chinshield 17. Tympanum (Ear/Eardrum)

13 Bearded Dragon *Pogona barbatus* (Green Valley, NSW).

14 Freshly dug and refilled nest of a Bearded Dragon *Pogona barbatus* on a fire trail at West Head, NSW.

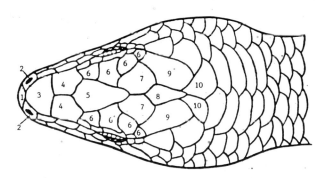

Scalation of a skink lizard's head (dorsal view).
(Pink Tongued Skink *Tiliqua gerrardi*).
1. Rostral 2. Nasal 3. Frontonasal 4. Prefrontal
5. Frontal 6. Supraocular 7. Frontoparietal
8. Interparietal 9. Parietal 10. Nuchal

Suborder Serpentes (Snakes)

In Australia there are seven recognised families of snake. They are:

1 Typhlopidae — Blind or Worm Snakes (found in all of mainland Australia).

2 Boidae — Pythons and Boas (found in most parts of mainland Australia).

3 Colubridae — Colubrid Snakes (found on the east coast and far northern Australia).

4 Acrochordidae — File Snakes (found in far northern Australia).

5 Elapidae — Front-fanged Venomous Land Snakes (found Australia wide).

6 Laticaudidae — Egg-laying Sea Snakes (found in northern Australian seas).

7 Hydrophiidae — Live-bearing Sea Snakes (found in all Australian seas, except south coast).

LOCOMOTION

Snakes move over land in four ways, and combinations of these. The first three methods are used by all snakes, whereas the fourth method is only used by a few species:

1 *Rectilinear movement* is when snakes move by minute muscular contractions of belly (ventral) muscles, which move in waves down the snake's body and enables the snake to move in a straight ahead or linear manner.

2 *Lateral undulation* is when snakes move by pushing the body forward in a series of curves which gain purchase from irregularities of the ground surface.

3 *Concertina movement* is used when the snake is moving over a smooth surface or climbing. One part of the snake's body is anchored to a given point, and the rest of the body is pulled to or pushed from this given point. The snake progresses as it finds new 'anchor points'.

4 The *sidewinder motion* involves the snake moving its body rapidly in a S-shape configuration and actually throwing its body sideways. Snaketracks left by snakes moving in this manner are often broken. This form of movement is usually only engaged when a snake is fleeing.

In order to effect locomotion, snakes usually possess between 200 and 400 vertebrae in their backbone. These are connected to each other by five individual joints, and most vertebrae support a pair of ribs which enclose the body cavity, excluding the ventral surface.

FEEDING

All snakes are predators. Within Australia, with the exception of Blind Snakes (Typhlopidae) which eat ants, termites and insects, all snakes feed exclusively on vertebrates and/or their eggs. In the wild snakes will only feed on live food captured by the snake. This food is usually killed before being eaten.

Prey is killed by constriction in many species (e.g. Boidae), while others rely on venom to immobilise prey (most Elapidae). Other species may simply eat food while it is still alive (most Colubridae and some Elapidae). Some species may feed using a combination of these three methods. Typical of the combination method are venomous species which will constrict (or hold) prey as well as biting it and injecting venom (e.g. Brown Snakes *Pseudonaja* spp.).

All species subdue prey by initially biting it. Constrictors will then wrap coils around the animal and kill it by constriction leading to asphyxiation. Venomous species bite prey to inject venom to it, and then may either hold on to the prey and wait for it to die, or release the prey to wander off and die, to be followed by the snake. Those snakes which eat food while it is still alive simply begin swallowing on biting its prey.

The elastic ligaments connecting snakes' jaws, referred to earlier, enable snakes to swallow prey, whole, of much greater diameter than their own head and body. When feeding, snakes will eat the prey in such a manner so as to lessen any resistance to movement of the animal down the snake's mouth and throat. This usually means that the animal is eaten head first. As the snake's jaws move over the body of the prey they become dislocated, with the upper and lower jaw being moved over the prey independently. The small sharp recurved teeth of snakes enables the snake to draw its head over its prey. Simultaneously the neck region undergoes a series of contractions also moving over the prey as necessary. Eventually the snake's head moves over the prey item and the prey then moves down the snake's body. Once the item is past the head, it moves rapidly down the body of the snake into the stomach, which is between half and two-thirds of the way down the snake's body. Muscular contractions of the snake often make the food item appear smaller once it is inside the snake's stomach. During feeding the windpipe is projected forward to assist in breathing. After feeding, the snake's jaws return to normal with the dislocated bones slotting back into place, though sometimes the snake must 'yawn' several times for this to happen.

The feeding process may take from sixty seconds to two hours depending on the proportional size of the food item, type of snake, how hungry the snake is, etc. For most snakes the average feeding time per food item is between two and twenty minutes.

Digestion of food may take up to two weeks; however, the time taken depends on the size of the food item, and the temperature of the snake after feeding. Usually complete digestion averages from four to six days for most snakes' meals. If a snake is too cool after feeding, so that digestion cannot take place, the food must be regurgitated or it might decay within the snake and kill it.

Undigestable items, such as bird feathers, are passed out in the faeces, or in some cases regurgitated after digestion of a meal.

The frequency with which snakes feed varies, but it can be safely stated that most snakes feed on average less than once a week. Some large snakes have been known to feed less than six times a year over several years. Although fasting periods between meals vary, some large snakes have been known to starve for more than twelve months without apparent ill-effects.

DEFENCE

Snakes typically are very secretive creatures and this is probably their major defence. Like all reptiles, when inactive, snakes will usually hide in places where they will not be detected. When approached most snakes will flee. Those which do not flee generally rely on their cryptic coloration to avoid detection.

When cornered a number of types of snake will adopt various means of defence. The obvious defence of snakes is to bite. All snakes possess teeth, and some species are venomous, even deadly to humans.

Some snakes will issue a warning to a would-be predator before biting. Some snakes will hiss loudly when harassed. Others will flatten out their head and neck region or even the whole body when provoked. In the Americas, Rattlesnakes (Crotalidae) 'rattle' their specially adapted tails when harassed.

14

Some snakes will lunge at an aggressor without attempting to bite.

A number of smaller elapids (e.g. *Simoselaps* spp.) will, when harassed, hold their body in a still manner and conceal the head beneath a coil. The Bandy Bandy *Vermicella annulata* will hold its body up in a series of stiff loops when harassed. The actual reason for this posture is not known.

MATING

Snakes of opposite sex cannot be identified on the basis of colour as they are usually the same, and apparently identify each other through scents produced by well-developed anal glands.

Sexual behaviour of most snakes is essentially similar. Because of the number of myths relating to snakes' mating habits, this subject is covered in detail here.

All snakes appear to be most sexually active during periods of falling air pressure, often associated with and preceding cold fronts, low pressure troughs, cyclones, etc. Temperature variation (seasonal or an unusual daily pattern) also appears to stimulate sexual activity in snakes. During the mating season snakes release scent from their anal glands which open into the cloaca. This scent is important for snakes finding one another, as they are usually solitary creatures.

Males of some species appear to have combat during mating seasons. The combat or 'dance' consists of the male snakes twining around one another and 'wrestling', usually keeping their heads and necks well above ground level. While wrestling, male snakes are usually oblivious to what goes on around them. Sometimes, but not always, male snakes will bite one another when in combat. It is not known if the male combat in snakes is territorial, in competition for a given female, or for both. More research is needed here.

Male snakes do not appear to engage in any real courtship prior to mounting the female. When mating commences the male snake will mount the female and align his body with that of the female so as to cover as much of the female's body as possible. The male snake will often flatten his body when doing this. The male will also caress the female snake with his head and 'chin' and with his tail. It is rare for a male snake to bite a female when mating.

The male snake will attempt to position his tail underneath the female's body and hopefully to make vent-to-vent contact. If the anal regions of the two snakes are not in alignment the male will edge his whole body backwards or forwards until their two vents are adjacent. Occasionally the male snake mounts the female the wrong way with his head over her tail and his tail adjacent to her head. When this happens the male will, after some frustration, reverse his positioning.

As the male snake gets aroused he will evert a hemipenis (penis equivalent), regardless of whether his vent is placed near the female's or not. Also the female snake might not always co-operate with the male snake in his lovemaking efforts, and may attempt to move away, usually with the male in hot pursuit.

Eventually, often after much effort, the male snake will insert a hemipenis into the vent of the female and copulation commences. The hemipenis then swells to an enormous degree and the male then becomes effectively connected to the female snake. During copulation the male snake will attempt to keep his body aligned over that

15 Brown Tree Snake *Boiga irregularis*, sub-adult (West Head, NSW).

16 Carpet Python/NSW Diamond Snake form *Morelia spilota spilota*, male (Hornsby Heights, NSW).

17, 18, 19, 20 Tiger Snake *Notechis scutatus*, feeding (Penrith, NSW).

15

of the female. The male snake will also twitch at various parts of his body, particularly the neck and tail regions. This twitching coincides with the arousal of the male, and as the male snake 'climaxes' the twitching becomes almost non-stop and apparently uncontrollable. Male snakes can have multiple climaxes.

If the male terminates copulation, he simply removes his hemipenis and moves away. If the female wishes to terminate copulation, she simply moves away dragging the male along with her, as he remains still connected by the hemipenis (which while swollen cannot be removed). The male, who presumably suffers discomfort in these circumstances, will attempt to realign his body on top of that of the female while copulating. If the female continues to move away from the male snake, dragging him with her, copulation will eventually be broken.

Some male snakes can spend literally days trying to mount female snakes, while others can mount and be copulating within sixty seconds. Copulation times range from a few seconds to several days, although for sperm transfer to be effective it is assumed that copulation of more than sixty seconds is generally required. Most snakes copulate for an average of two to three hours at a time, to effect a successful conception.

MAJOR DIAGNOSTIC FEATURES

Many snakes are superficially similar in appearance so the use of diagnostic features to identify given species becomes more important. Major features of snakes used to identify different types include:

1 Scalation, head and body.

2 Morphology.

3 Colour

4 Size.

21   Brown Snake *Pseudonaja textilis* about to shed skin — note clouded eye (Cobar, NSW).

22   Male Death Adder *Acanthophis antarcticus* about to shed skin. Note dull coloration.

23   Male Death Adder *Acanthophis antarcticus* immediately after sloughing, with shed skin.

24   Carpet Python *Morelia spilota macropsila*, female, head (Bundaberg, QLD).

25   Ant-hill Python *Bothrochilus perthensis*, (?) female, head
(Katherine, NT).

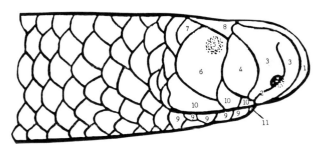

Scalation of a Blind Snake's head (lateral view).
(*Ramphotyphlops nigrescens*)
1. Rostral   2. Nasal Cleft   3. Nasal   4. Preocular   5. Eye
spot   6. Ocular   7. Parietal   8. Supraocular   9. Infralabial
10. Supralabial   11. Mental

26   Death Adder *Acanthophis antarcticus* male, grey phase
(Glenbrook, NSW).

27   Golden-crowned Snake *Cacophis squamulosus* male (West
Head, NSW).

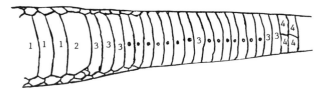

Scalation of a snake's tail region (ventral surface)
(Death Adder *Acanthophis antarcticus*).
1. Ventral   2. Anal   3. Subcaudal (single)   4. Subcaudal
(divided or paired) (each pair is counted as one)

28 Stimson's Python *Bothrochilus stimsoni* in defensive posture hiding its head among its tight coils (Barkly Tableland, NT).

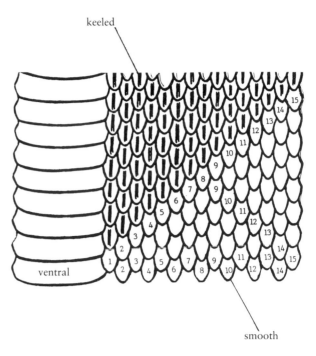

keeled

ventral

smooth

Counting mid body scale rows in a snake (generalised snake)

29 Death Adders *Acanthophis antarcticus* copulating (NSW).

30 Death Adders *Acanthophis antarcticus* copulating, showing connection point at the base of the tails (NSW).

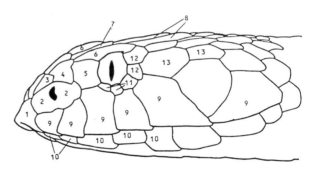

Scalation of a snake's head (lateral view).
(Death Adder *Acanthophis antarcticus*).
1. Rostral 2. Nasal 3. Internasal 4. Prefrontal
5. Preocular 6. Supraocular 7. Frontal 8. Parietal
9. Supralabial 10. Infralabial 11. Subocular
12. Postocular 13. Temporal

FALLACIES

Since the beginning of time reptiles, particularly snakes, have been worshipped, loved or hated, and there are probably more fallacies and myths concerning reptiles than any other type of animal. Corrections to some common fallacies are given here.

* A snake's body is not slimy, but is dry to touch and, unlike most animals, they are effectively odourless to humans.
* Hoop snakes — snakes which can grab their tails and roll down hills — do not exist. Snakes that can crack like whips (and live) are also mythical.

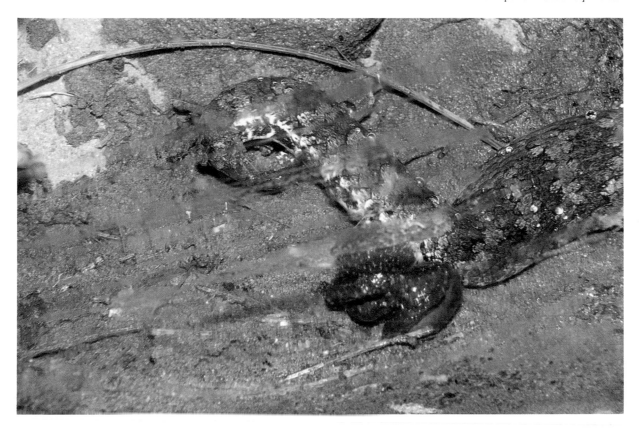

31 Erect hemipene of Death Adder *Acanthophis antarcticus* immediately after breaking copulation. Note the bi-lobed nature of this organ, and the shape of each lobe. These essentially lock the Death Adders together when copulating. Snakes and lizards possess two hemipenes. (Photo taken through glass)

- Snakes do not avenge the death of mates, nor do they have nests (the only exception being the King Cobra *Ophiophagus hannah*, which does have nests). Snakes are usually solitary animals and usually only come together to mate. Snakes do not stay in pairs for life, like some other animals, and apparently have no concern for the welfare of one another.
- Snakes are not charmed by music. They lack external ears, are effectively deaf to airborne sounds and are charmed by movements of the charmer's bodies and not their music.
- Snakes do not milk cows. It is physically impossible and snakes do not drink milk. If snakes are offered a choice between water and milk, water will be taken every time.
- No snakebite or goanna bite leaves an annually recurring sore. If they did, some herpetologists would be permanently covered with sores!
- Snakes do not sting with their tongues. The tongue is merely a sensory organ.
- A Death Adder does not sting with its tail. The spine on the end of the tail is only a soft modified scale.
- Snakes are not habitually aggressive or evil and will never go out of their way to attack someone. They only attack if provoked.
- The Bandy-Bandy is not deadly, nor does its bite cause a fit for every ring on its body. The Bandy-Bandy, though weakly venomous, is effectively harmless to humans.
- No snakes swallow their young for protection. Fully-formed snakes occasionally found in freshly killed snakes are either young in the oviducts awaiting birth or they might have been eaten as a meal.
- Milking snakes does not make them harmless. Snakes' venom glands are similar in form to our own salivary glands, which means that they are never dry, and always carry enough venom to cause a lethal bite (if the snake is dangerous), even after milking. (The only way to make a deadly snake harmless is by surgical removal of the venom glands — a complicated procedure.)
- Reptiles do die at sunset, but they also die at any other time, depending on when the reptile was killed. Nerves of reptiles are often active long after death, and reptile hearts have been known to beat for days after death.
- Removal of a snake's fangs does not make them harmless. Snakes regularly shed fangs which are replaced by new fangs that grow in their place.
- Goanna oil does not seep through glass, nor does it contain any other magical quality.
- No snake has, or can develop, legs. Occasionally when killed a male snake's hemipenes will protrude from the hind body, appearing to be leg-like. Pythons and other reptiles have spur-like scales on the hind body which may look leg-like. They have a purely sexual role. The 'spurs' are homologous to claws, and are believed to have evolved from the remnants of the hind limbs.
- No snake hypnotises its prey.
- No snake sheds its tail like a lizard and minute snakes, which break into pieces and join back up again, are non-existent.
- A snake will not commit suicide by biting itself. All snakes are immune to their own venom. (Snakes will commonly bite themselves when dying, but the venom does not cause death. For example, snakes hit by car tyres will commonly bite themselves in a reflex action but the cause of death is the compression of the body by the weight of the vehicle.)

32 Death Adder *Acanthophis antarcticus* male, grey phase (West Head, NSW).

33 Death Adder *Acanthophis antarcticus* male, red phase (Cottage Point, NSW).

SNAKEBITE AND ITS TREATMENT

Snakebite should always be taken seriously. Most snakebite deaths in Australia result from the victim not taking the bite seriously or not treating the bite properly.

Even if a snakebite is only suspected, or there was a chance that the snakebite was from a harmless variety, it is better to go to a hospital rather than wait for symptoms to occur, at which stage it may be too late to save the victim. Hospitals store anti-venoms as well as other valuable aids to the survival from snakebite. Snake venoms give varying symptoms to different people and often their effects are not felt for more than twenty-four hours. Snake bites are not always visible and it is dangerous to assume that when no punctures are visible that no bite has occurred.

A snakebite victim should never be given a sedative such as alcohol or a stimulant such as strychnine or coffee, nor should they be allowed to move unless totally necessary, as all of these will increase the rate at which venom reaches the heart. If one is bitten while alone, *walk* — do not run — for help.

Australia's deadly snakes have neurotoxins as the principal deadly component of their venoms which means that they attack the voluntary muscles and the nervous system. Death is usually the result of suffocation caused by blockage of the respiratory passage by some means.

Other components of snake venom include: anticoagulants, which prevent blood clotting; cytolysins, which destroy blood and other cells with which they come into contact; haemolysins, which destroy red blood cells; haemorrhagins, which destroy linings of blood vessels, permitting blood to escape into adjacent body tissue; and thrombase, which causes blood clotting within vessels.

One should remember that nowadays death from snakebite is very rare, and death from any snakebite within three hours is virtually unknown. Snakebite treatment has changed significantly over recent years, particularly with the development of polyvalent antivenoms. These are antivenoms which are useful against all snakes found in a given state or country. The treatment most commonly recommended by experts now in the event of snakebite is the following:

1 Apply a firm broad constrictive bandage at the site of the bite.

2 Immobilise the limb and, if possible, the whole person.

3 Reassure the patient, as stress is an unwanted harmful factor.

4 Get medical treatment, in particular antivenom if required. If possible alert the local hospital of the impending arrival of the patient.

Things not to do about snakebite and its treatment:

- Do not cut the wound, or use Condy's crystals (potassium permanganate) in treatment.
- Do not use a torniquet, unless you do not have access to something that can be used as a broad constrictive bandage.
- Do not suck the wound.
- Never amputate the limb as the risk of serious infection will be greater than that from the actual bite.
- Do not try to catch or kill the snake for identification purposes, as there is the risk of further bites. The use of polyvalent snake antivenoms means that it is not crucial to identify the snake.
- Do not wash the wound, as venom remaining on the skin surface may be identified with CSL (Commonwealth Serum Laboratories) kits. Then the appropriate specific antivenom can be used in preference to the polyvalent.
- Do not restrict free chest movement or breathing activity by the patient.

PART 2 AUSTRALIAN FROGS AND REPTILES DESCRIBED

ABOUT THE DESCRIPTIONS

I HAVE DEALT with a selection of Australia's reptiles and frogs in systematic order, grouping given species in terms of their phylogenetic (evolutionary) relationships. As some Australian species are not included, this book should not be regarded as a strict or definitive identification manual.

The species included in the book have been selected on the following basis. Those most likely to be encountered, those best known or those of special interest to herpetologists have been included. I have tended to exclude smaller, more obscure species, and species that are similar to one another, and have only provided descriptions of a few of these types.

The species which tend to be kept in captivity the most have the most detailed descriptions.

Some reptiles have been scientifically described but not given a common or colloquial name. I have not attempted to coin names for these species, so they are listed in this book with scientific names only.

'The common names given here are those most widely in use. In a few cases, where a species is widely known by more than one common name alternatives may be given. However, I have not attempted to list all the obscure names that any given species may occasionally be called.'

For each family from which I have selected species, I have provided a description of the general characteristics of each family, before going into the given species in detail.

I have minimised descriptive information on the appearance of each species, mostly relying on the photos to complement the description. Where a given species is variable in form or colour, I have given some added descriptive information. Unless otherwise stated all measurements quoted are of adult specimens and from the tip of the snout to the end of the tail (as opposed to snout-to-vent measurements).

Most reptiles and frogs typically produce a number of offspring at a time. Occasionally as a result of some anomaly these types may ovulate to produce only one or two young or eggs, being an abnormal pattern for the species. When discussing numbers of eggs or live young produced by these species, I have disregarded these 'abnormal cases'.

Because snakes tend to look more alike than other reptiles and frogs, I have included basic scale counts for each species described; this is a useful aid to identification. Also the fact that some snakes are potentially dangerous makes more positive identification necessary.

For each species shown here I have provided information relating to known distribution. I have also provided known information about the biology of these species, much of it the result of my own research over the last twenty years. There is a relative paucity of information on many species, due to the relative lack of research undertaken on Australia's reptiles and frogs.

CLASS AMPHIBIA

Order Salientia (Frogs and Toads)

Family Bufonidae (Toads)

These 'true toads' do not occur naturally in Australia, although they are abundant in most other parts of the world. Toads are characterised by a warty dry skin, and poison glands, usually highly visible, on the back of the head and elsewhere. Reproductive modes and lifecycles of different species of toad vary considerably.

Cane Toad *Bufo marinus*

CANE OR MARINE TOAD
Bufo marinus (Linnaeus, 1758)
Figs 35, 36

The only 'true toad' found in Australia, the Cane Toad was introduced into north-eastern Australia around 1935 from Hawaii and South America, in order to control the pest Sugar Cane Beetle *Dermolepida albohirtum*. The toad itself became a major pest and its distribution grows annually. Although found in many habitats, the Cane Toad's highest population densities occur near human habitation. Currently found all over Queensland and adjacent areas, it is spreading south in New South Wales and west across the Northern Territory at the rate of about 30 km per year, and it might eventually be found in most parts of Australia where habitats are favourable unless immediate steps are taken to stop its expansion.

The toads are a major pest for several reasons. Their tadpoles eat other frogs' tadpoles, and because the toads will breed in all available fresh or brackish water bodies, other species of frog die out when toads become established in their native areas. The toads also lay thousands of eggs at a time, thereby allowing their range to increase at an alarming rate. A single female is capable of laying some 35,000 eggs per year.

34 Axolotl *Ambystoma mexicanum* (captive specimen).

35 Cane Toad *Bufo marinus* (Bundaberg, QLD).

36 Cane Toad *Bufo marinus*, head (Grafton, NSW).

Most animals that are small enough to fit into the mouth of a toad will be eaten. Species that might eat the toads will be poisoned. Therefore the toads wipe out all frogs, frog-eating reptiles and birds, and most other vertebrate wildlife, either directly or indirectly, when they move into an area.

Cane Toads are large and ugly in appearance, up to 20 cm in length, with large, highly visible poison glands (the parotid glands) located at the back of the head and elsewhere.

The call of the Cane Toad is an unmistakable mechanical-sounding 'pop-pop-pop'.

Family Myobatrachidae (Southern Frogs)

Commonly called 'Southern Frogs', this family is only found in Australia and nearby areas. Similar related families are found in South Africa and South America. Most species of Australian frog belong to this family. Most species are relatively heavily built ground-dwellers and/or burrowers, although a huge diversity of forms and habits occur. Reproductive modes vary from egg–tadpole–frog stages in some species to the direct egg–frog mode in other species. In some species the larvae develop within the frog's digestive system.

COMMON EASTERN FROGLET
Crinia signifera (Girard, 1853)
Figs 37, 38

This small, 3-cm frog is highly variable in colour pattern, even within a given population. Dorsally it ranges from black in colour, through all shades of brown and grey. The pattern might be plain or with varying numbers of thick longitudinal stripes, again of varying colour. Some specimens are warty, while others have smooth skins.

This frog is found throughout south-eastern Australia and Tasmania, and similar, almost undistinguishable species are found in most parts of Australia. Complicating things further, several species of *Crinia* may live in the same place, and only experts can distinguish their various calls.

These small frogs occupy all habitats, and are common in very built-up areas, such as Sydney, Melbourne and

37   Common Eastern Froglet *Crinia signifera* (Wentworth Falls, NSW).

38   Common Eastern Froglet *Crinia signifera* (Ourimbah, NSW).

Brisbane. Froglets (*Crinia*) shelter under debris near semi-permanent and permanent watercourses, swamps, damp street gutters, etc.

They breed all year round, and feed on very small insects. The eggs are usually deposited in several places and invariably end up attaching themselves to partially submerged vegetation, and other things.

The call of *Crinia signifera* is a distinctive 'crick, crick, crick' sound. Closely related species in this genus have both similar and distinctly different calls, depending on species and locality.

GIANT BURROWING FROG
Heleioporus australiacus (Shaw and Nodder, 1795)
Figs 4, 39–41, 535

This large distinctive frog attains about 10 cm in length and is found from about Newcastle in New South Wales, along the coast and ranges south to the ranges east of Melbourne, Victoria.

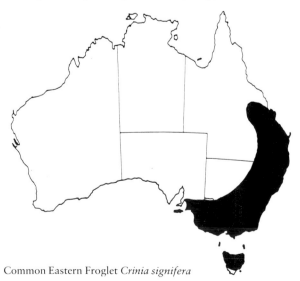

Common Eastern Froglet *Crinia signifera*

39 Giant Burrowing Frog *Helioporus australiacus* male (West Head, NSW).

40 Giant Burrowing Frog *Helioporus australiacus* female (West Head, NSW). Note thinner forearms and lack of black spines on toes.

41 Giant Burrowing Frog *Helioporus australiacus* head (Mount White, NSW).

It is usually only found around sandstone plateaux, where it breeds through summer and autumn in small permanent creeks. Mating and egg-laying often occurs in disused yabbie holes along creek banks. The spawn consists of up to 400 large eggs in a frothy mass.

The male of this species has much thicker forearms than the female, and black spikes on the end of some fingers, to aid his grip of the female when mating.

The Giant Burrowing Frog is active all year, and appears to move great distances from breeding areas when apparently foraging for food at night. This species may even be found moving about on nights too hot and dry for most other frogs. The tadpoles are distinctly large and black.

The call of this frog is an owl-like 'oo-oo-oo'.

Giant Burrowing Frog *Helioporus australiacus*

FLETCHER'S FROG
Lechriodus fletcheri (Boulenger, 1890)
Fig. 42

This moderately built frog attains 5 cm. Common in rainforest areas along the New South Wales and south-east Queensland coast, it may also occur near Cairns, North Queensland.

This frog occurs alongside the Great Barred Frogs (genus *Mixophyes*) and is commonly mistaken for being

Fletcher's Frog *Lechriodus fletcheri*

immature *Mixophyes* specimens. Fletcher's Frog does not live in areas heavily disturbed by man adjacent to its habitat. Fletcher's Frog is a summer breeder in small creeks and ponds within rainforest. It feeds on insects, and other small animals, and its tadpoles are cannibalistic.

The call is a 'g-a-r-u-p' lasting about a second, repeated every few seconds.

MARBLED FROG
Limnodynastes convexiusculus (Macleay, 1877)
Fig. 43

Found throughout coastal and near-coastal areas of the Northern Territory and North Queensland, this frog attains about 5 cm.

Its preferred habitat is low-lying swampy areas in savannah woodlands, where specimens breed in thick grass verges and adjacent to other cover. Large numbers of this species are often found within a relatively small area.

This insectivorous species breeds in the northern wet season after the first major rains. Its call is a repeated high-pitched 'plonk'.

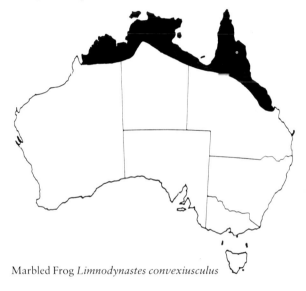

Marbled Frog *Limnodynastes convexiusculus*

EASTERN BANJO FROG
Limnodynastes dumerilii Peters, 1863
Figs 44–47

A stocky ground-dweller which attains 7 cm, this frog is very variable in dorsal colour, ranging from grey to brown or even reddish.

This species is found on the coast, ranges and near slopes of south-eastern Australia. Similar related Banjo Frog species are found in most other parts of Australia.

The frog usually is found close to dams and swamps, in all habitats within its range, rather than flowing creeks, which it seems to avoid. Because this species hides under submerged banks and underground in mud during the day, it is effectively impossible to locate during daylight and is commonly found at night foraging around swamps and dams for insects.

It is called the Banjo Frog, or 'Pobblebonk', because of its distinctive call, which sounds like a series of 'plonks' or the plucking of banjo strings. Males usually call from very well-concealed points, commonly submerged air pockets.

42   Fletcher's Frog *Lechriodus fletcheri* (Wyong, NSW).

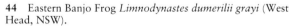

43   Marbled frog *Limnodynastes convexiusculus* (Jabiru, NT).

44   Eastern Banjo Frog *Limnodynastes dumerilii grayi* (West Head, NSW).

45 Eastern Banjo Frog *Limnodynastes dumerilii grayi* sub adult (Wyong, NSW).

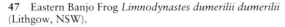

46 Eastern Banjo Frog *Limnodynastes dumerilii dumerilii* (Lithgow, NSW).

47 Eastern Banjo Frog *Limnodynastes dumerilii dumerilii* (Lithgow, NSW).

When eggs are laid the female beats the eggs into a floating frothy mass, where they remain until hatching into tadpoles a few days later. It breeds in the warmer months.

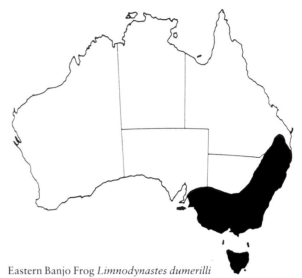

Eastern Banjo Frog *Limnodynastes dumerilli*

GIANT BANJO FROG
Limnodynastes interioris Fry, 1913
Fig. 48

Closely related to the Eastern Banjo Frog, the Giant Banjo Frog grows to 10 cm and is distinguishable by its larger size and usually brighter yellow belly. It is found on the western slopes and plains of New South Wales.

This frog feeds on small arthropods and breeds throughout the year. Otherwise its call, mating and other habits are typical of all Banjo Frogs.

48 Giant Banjo Frog *Limnodynastes interioris* (Dubbo, NSW).

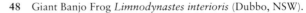

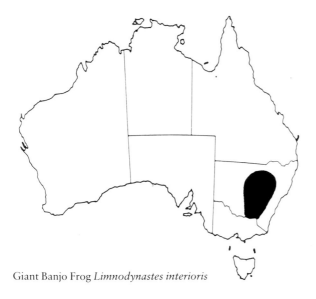

Giant Banjo Frog *Limnodynastes interioris*

49   Ornate Burrowing Frog *Limnodynastes ornatus* (Yamba, NSW).

ORNATE BURROWING FROG
Limnodynastes ornatus (Gray, 1842)
Fig. 49

This small frog may occasionally reach 5 cm in length, although most adults are usually smaller than this. The Ornate Burrowing Frog is found in all habitats in the eastern and northern parts of Australia, from deserts to rainforests.

This insectivorous frog is usually seen after rain, commonly feeding near roadsides, where insects appear to congregate. When breeding this frog may be found in large numbers around permanent and semi-permanent waterholes and ponds. The Ornate Burrowing Frog has a very rapid egg–tadpole frog cycle, an adaptation to beat the drying out of the breeding waterholes. The call is a rapidly repeated nasal 'unk'.

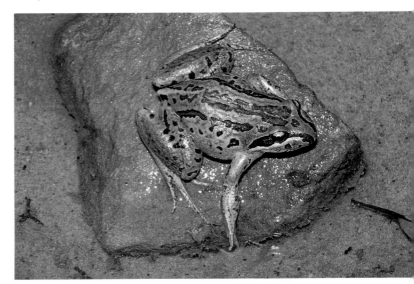

50   Brown-striped Frog *Limnodynastes peronii*, light brown (West Head, NSW).

51   Brown-striped Frog *Limnodynastes peronii*, dark brown (Ourimbah, NSW).

Ornate Burrowing Frog *Limnodynastes ornatus*

BROWN-STRIPED FROG
Limnodynastes peronii (Dumeril and Bibron, 1841)
Figs 2, 3, 50, 51

This common frog of 7 cm is found along the coast and ranges of eastern Australia and Tasmania.

52 Spotted Grass Frog *Limnodynastes tasmaniensis*, with vertebral stripe (Wyong, NSW).

53 Spotted Grass Frog *Limnodynastes tasmaniensis*, without vertebral stripe (Dubbo, NSW).

54 Northern Banjo Frog *Limnodynastes terraereginae* (Yamba, NSW).

It breeds in the warmer months of the year, laying eggs in a floating mass, in swamps, dams, marshes and ponds.

This frog is even common in built-up areas. Its call is a distinctive 'toc' sound, repeated every few seconds. Males will usually call from underneath something, such as a submerged bank, vegetation or even an already-laid floating egg mass.

The tadpoles are large and black, and attain 6.5 cm in length before metamorphosing.

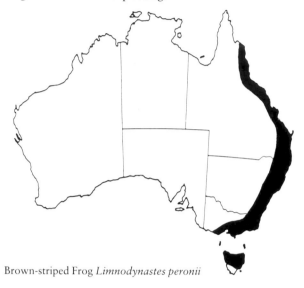

Brown-striped Frog *Limnodynastes peronii*

SPOTTED GRASS FROG
Limnodynastes tasmaniensis Gunther, 1858
Figs 52, 53

Found in all habitats from the coastal forests to inland deserts, the Spotted Grass Frog is found throughout the eastern half of Australia and Tasmania. An introduced isolated population is found in north-western Western Australia, near Kunnanurra.

This frog is highly variable in colour pattern, even between individuals from the same place. Dorsally some specimens have olive or brown mottling, others have distinctive blotches, others are unmarked, whilst many have a white, yellow, orange or red stripe running down the middle of the back. It attains 5 cm.

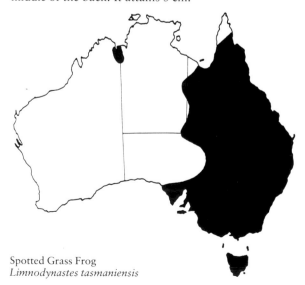

Spotted Grass Frog
Limnodynastes tasmaniensis

This frog is usually caught sheltering under debris near grassy swamps, dams and lagoons, its preferred habitat. Sometimes massive numbers may congregate in a very small area. Breeding can occur at any time of year.

The call is a rapid 'uk-uk-uk' in New South Wales and further north, and a single 'click' sound sometimes repeated, in southern locations elsewhere.

NORTHERN BANJO FROG
Limnodynastes terraereginae Fry, 1915
Figs 6, 54

Closely related to other Banjo Frogs, this 7.5-cm species occurs along the coast and ranges of Queensland and northern New South Wales.

This species has been decimated by the Cane Toad *Bufo marinus*, which seems to thrive in the same habitats as this frog. The call of this frog is shorter and more highly pitched than that of the other Banjo Frogs.

Its habits are otherwise similar to those of the Eastern Banjo Frog *Limnodynastes dumerilii*.

Northern Banjo Frog
Limnodynastes terraereginae

BARRED FROG
Mixophyes balbus Straughan, 1968
Fig. 55

Barred Frog *Mixophyes balbus*

55   Barred Frog *Mixophyes balbus* (Wattagan Ranges, NSW).

56   Great Barred Frog *Mixophyes fasciolatus* (Mount Glorious, QLD).

Found along the coast and near ranges of New South Wales and adjoining areas, this large frog is restricted to rainforests and similar habitats within its range.

Like all 'Barred Frogs' it has a distinctive appearance. It reaches 8 cm in length.

This ground-dwelling frog eats all arthropods and vertebrates that will fit into its mouth, including other frogs and small snakes.

The female is the larger sex and lays her eggs on vegetation adjacent to fast-flowing rainforest creeks. The eggs are washed into the creeks by rain where they hatch and the tadpoles subsequently develop into frogs.

The call is a short grating 'trill' of several pulses.

GREAT BARRED FROG
Mixophyes fasciolatus Gunther, 1864
Fig. 56

This species is very closely related to the Barred Frog *Mixophyes balbus*. The Great Barred Frog is found in

57   Painted Burrowing Frog *Neobatrachus sudelli*, male (Amphitheatre, VIC).

58   Northern Holy Cross Frog *Notaden nichollsi* (Barkley Tableland, NT).

Great Barred Frog *Mixophyes fasciolatus*

rainforests of the coast and near ranges of New South Wales, and Queensland south of Bundaberg. It reaches 8 cm in length.

This ground-dweller appears to have held its own against the Cane Toad *Bufo marinus*, which appears to have decimated most other frog species in the areas it invades.

The Great Barred Frog is very 'jumpy' and highly strung, and is a regular winner of frog-jumping competitions.

The biology of this frog is essentially similar to that of the preceding species, although its call is a deep harsh 'Wack'.

PAINTED BURROWING FROG
Neobatrachus sudelli (Lamb, 1911)
Figs 57, 564–571

This stoutly built, 4-cm frog is found throughout inland Victoria and New South Wales and adjacent parts of Queensland and South Australia. Similar related species are found in most parts of mainland Australia. It is usually only found after rain, when it breeds in temporary pools, swamps and small creeks.

The Painted Burrowing Frog usually has a warty skin, which becomes spinose in breeding males. On each hind foot is a large black, hard, metatarsal tubercle, which aids in burrowing in soft sand and mud. When burrowing, this frog burrows hind first by using its feet as shovels. When descending, it rotates its body (illustrated). All Burrowing Frogs tend to burrow in this manner. During dry seasons this frog remains underground or under deep cover, awaiting the next rains.

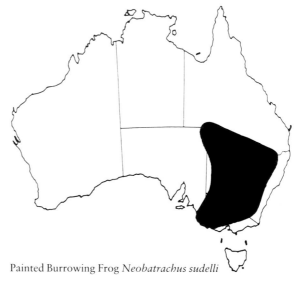

Painted Burrowing Frog *Neobatrachus sudelli*

NORTHERN HOLY CROSS FROG
Notaden nichollsi Parker, 1940
Fig. 58

This medium-sized frog grows to 6 cm and is found in most arid parts of the northern half of Australia. It is usually found in red or black soil areas after heavy rain, and is one of the dominant frog species where it occurs.

Related species are found in most of the drier parts of Australia. Although a waterholding frog, the Northern Holy Cross Frog is not used by the Aborigines as a water source in droughts because it is small and exudes a thick smelly substance when agitated, which when dry becomes strong and elastic.

It breeds in waterholes formed after the heavy rain. Its tadpoles develop rapidly to metamorphosis.

The Northern Holy Cross Frog has a call 'oo-oo' which sounds pigeon-like.

59 Red-crowned Toadlet *Pseudophryne australis* (Terrey Hills, NSW).

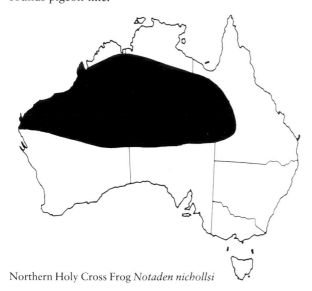

Northern Holy Cross Frog *Notaden nichollsi*

RED-CROWNED TOADLET
Pseudophryne australis (Gray, 1835)
Fig. 59

This distinctive small frog is found only in the sandstone hill country to the north, south and west of Sydney, within a radius of about 150 km.

It averages 3 cm in length and is usually found around seepages on hillsides, associated wet rock crevices and under nearby rocks. In these situations this species often occurs in colonies numbering up to fifty individuals, where they can often be heard calling to one another.

The eggs laid are firm, and deposited in moist crevices or next to small ponds, often in a crude nest, where the young tadpoles live after hatching. It is assumed that from egg to young frog takes about three months, although young toadlets are rarely seen.

The call of this frog is a harsh squelch sound. However, as this and most other species of *Pseudophryne* have lost the outer and part of the inner ear, just how they actually hear airborne sounds is a mystery.

60 Brown Toadlet *Pseudophryne bibronii* (Lithgow, NSW).

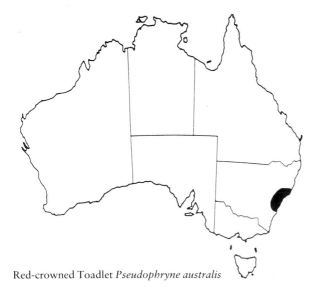

Red-crowned Toadlet *Pseudophryne australis*

BROWN TOADLET
Pseudophryne bibronii Gunther, 1858
Fig. 60

A close relative of the Red-crowned Toadlet, this species also grows to about 3 cm in length.

The Brown Toadlet is found throughout most of south-eastern Australia and Tasmania. It occupies most habitats, but is not found in sandstone areas around Sydney.

It lives under rocks, logs and other debris near semi-permanent streams and swamps, where they breed when it rains. Although known to breed at all times of the year this species most often breeds in the cold winter months, when it has been recorded as calling in temperatures as low as 4° C. This frog lives on flies, mosquitos and similar insects.

The call is a short, reverberating, upwardly inflected 'ark'.

61   Keferstein's Frog *Uperoleia laevigata* (Glenorie, NSW).

62   *Uperoleia* sp. (Boggabri, NSW).

KEFERSTEIN'S FROG
Uperoleia laevigata Keferstein, 1867
Fig. 61

Little is known about this species, a small 2.5-cm frog from coastal and near-coastal New South Wales. A number of similar frogs from the genus *Uperoileia* are found in most parts of Australia (*Laevigata* species complex), and all might eventually be reclassified and split into more new species. Differentiating between one species in this group and another is exceedingly difficult.

These frogs are usually found in the vicinity of man-made dams, semi-permanent swamps and so forth. They breed in spring and summer. Eggs are usually laid, singly and in small clumps, at the bottom of shallow water.

The call is a short harsh 'ahk'.

Keferstein's Frog *Uperoleia laevigata*

Uperoleia orientalis (Parker, 1940)
Fig. 63

This frog is easily distinguished from *Uperoileia laevigata* (see illustration). This slightly larger species grows to about 3 cm and is only found in the tropical part of the Northern Territory.

A ground-dwelling species, it is found in large numbers breeding around swamps and flooded grasslands during the wet season. *Uperoleia orientalis* is generally nocturnal.

Brown Toadlet *Pseudophryne bibroni*

Uperoleia orientalis

Family Hylidae (Tree Frogs)

This family includes all the well-known tree frogs. Most species have large suction pads which enable efficient climbing of most things, including glass windows. Species range in form from agile climbers to ground-dwellers with reduced suction pads, and even burrowers which only have remnant suction pads. Most species are found in warm and moist areas, but this family is found Australia wide. Tree frogs occur throughout most parts of the world. Most, if not all, species lay eggs, which hatch into free-swimming tadpoles, which subsequently turn into frogs.

NORTHERN WATER-HOLDING FROG
Cyclorana australis (Gray, 1842)
Fig. 64

This large, robust frog attains 10 cm in length. Although a well-built burrowing frog, it is actually more closely related to other tree frogs (with which it is sometimes classified at family level) than to other Australian burrowing frogs. Like most burrowing frogs, it appears to hide underground during dry seasons, in a cocoon-like set-up.

This frog got its name from the fact that Aboriginals in desert areas used it as a source of water during droughts. They dug up water-engorged frogs in dried-up ponds, placed the rear of a frog in their mouth and squeezed water from the frog.

It is common throughout north-western and nearby parts of Australia. Found in most habitats, it is most commonly found crossing roads after tropical thunderstorms. This species has cannibalistic tendencies, and appears usually to breed in dams and semi-permanent waterholes, usually during summer. Its call is a distinctive and loud 'honk, honk'. The colour of this species changes depending on the physical state of the frog, its age, and other factors. Its colour ranges from olive, brown, grey and even pink above.

Northern Water-holding Frog
Cyclorana australis

NORTH-EASTERN WATER-HOLDING FROG
Cyclorana novaehollandiae Steindachner, 1867
Figs 65–67

This is a 10-cm frog which is similar in most respects to *Cyclorana australis*, from which it can only be easily dif-

63   *Uperoleia orientalis* (Humpty Doo, NT).

64   Northern Water-holding Frog *Cyclorana australis* (Arnhem Hwy, NT).

65   North-eastern Water-holding Frog *Cyclorana novaehollandiae* (Moonie, QLD).

66   North-eastern Water-holding Frog *Cyclorana novaehollandiae* (Charters Towers, QLD).

67   North-eastern Water-holding Frog *Cyclorana novaehollandiae* (Charters Towers, QLD).

68   Striped Burrowing Frog *Litoria alboguttata* (Moonie, QLD).

ferentiated by a lack of patterning on the back of the thighs. It is, however, found in north-eastern Australia only (as opposed to north-west).

This species is an opportunistic feeder, feeding mainly on insects, but will eat geckoes if available. In east coastal rainforest areas this species may be active at all times of the year because of the lack of drought conditions.

This frog can sit in ponds of water with only its nostrils showing. The ability to do this is an important means of avoiding predators.

It breeds during the northern wet season.

North-eastern Water-holding Frog
Cyclorana novaehollandiae

STRIPED BURROWING FROG
Litoria alboguttata (Gunther, 1867)
Fig. 68

This frog's dorsal colour ranges from brown to green, with or without stripe(s) or blotches. The Striped Burrowing Frog attains 6 cm and is found throughout most parts of Queensland, northern inland New South Wales, and adjacent to the Gulf of Carpentaria.

This frog is commonly found along floodplains, and around semi-permanent dams and waterholes. It has a relatively strong tolerance for brackish environments. Feeding on small arthropods, this species is active during the day in colder weather.

Striped Burrowing Frog *Litoria alboguttata*

This frog appears to be closely related to the Water-holding Frogs, genus *Cyclorana*, although more research is needed here.

The call is a loud and rapid quacking sound.

GREEN AND GOLDEN BELL FROG
Litoria aurea (Lesson, 1829)
Fig. 69

This frog grows up to 9 cm long, although males are usually somewhat smaller. As its name suggests, this species is green and gold in colour, with the gold consisting of variably shaped blotches. Although a 'tree frog', the Green and Golden Bell Frog only has small pads on its fingers and toes.

Essentially a pond-dwelling species, this voracious cannibalistic frog is most common around swamps and creeks with plenty of bullrushes, in which it shelters. When foraging for food at night, this species rarely strays far from its preferred habitat. During the summer breeding season it may also be diurnal. Eggs are deposited amongst loose floating vegetation, and tadpoles take about 10–12 weeks to metamorphose.

The call is a deep droning sound followed by about four short 'Crawk' sounds.

69 Green and Golden Bell Frog *Litoria aurea* male (Wyong, NSW).

70 Green Tree Frog *Litoria caerulea* (Ourimbah, NSW).

This species is very hardy in captivity, with larger specimens regularly feeding on mice.

Green and Golden Bell Frog *Litoria aurea*

GREEN OR WHITE'S TREE FROG
Litoria caerulea (White, 1790)
Figs 5, 7, 70

Probably Australia's best known frog, the Green Tree Frog is found throughout the eastern and northern half of Australia. Adults average 10 cm in length in most areas. The largest specimens of this species that I have seen have been from around Charters Towers, Queensland, where 15 cm is a common adult body length.

Although found in all types of habitat, this frog is usually found around human habitation, such as toilets, water tanks, etc. A summer and wet season breeder, Green Tree Frogs breed in old (and new) water tanks, drainage systems and semi-permanent swamps. The distinct dark-green tadpoles grow to about 9 cm before metamorphasing.

The call is a distinctive 'Wark-wark-wark'. In Queensland this is one of the few frog species that have not been too adversely affected by the invasion of the Cane Toads.

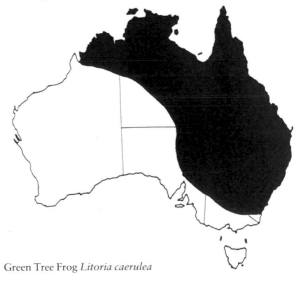

Green Tree Frog *Litoria caerulea*

71   Red-eyed Green Tree Frog *Litoria chloris* (Ourimbah, NSW).

72   Red-eyed Green Tree Frogs *Litoria chloris*, male and female. Female is the larger frog. (Wyong, NSW).

73   Blue Mountains Tree Frog *Litoria citropa* (Wattagan Ranges, NSW).

RED-EYED GREEN TREE FROG
Litoria chloris (Boulenger, 1893)
Figs 71, 72

A spectacular medium-to-large frog of 6 cm, the Red-eyed Green Tree Frog is found in rainforest and adjacent habitats along the east coast of New South Wales and Queensland. Most specimens are found crossing roads on rainy nights.

It breeds in permanent and semi-permanent waterholes, swamps and similar places, apparently avoiding fast-flowing streams. The breeding season is throughout the warmer months but apparently peaks in early December when this frog seems to be most abundant.

The sound of large numbers of this frog calling in rainforests after rain at night can be very loud indeed. The call is a series of long 'moans' sometimes followed by a soft 'trill'.

Red-eyed Green Tree Frog *Litoria chloris*

BLUE MOUNTAINS TREE FROG
Litoria citropa (Dumeril and Bibron, 1841)
Fig. 73

This nicely coloured frog reaches 6 cm, and is found along the coast and ranges of south-east New South Wales and Victoria.

Blue Mountains Tree Frog *Litoria citropa*

This frog is rarely found in large population densities. It is usually located in rock crevices, or under rocks near small permanent and semi-permanent creeks. Occasionally adult specimens are found considerable distances from any possible breeding sites, usually in winter. The Blue Mountains Tree Frog breeds in early spring.

The call is a sharp scream followed by a trill.

BLEATING TREE FROG
Litoria dentata (Keferstein, 1868)
Fig. 74

A small 4-cm frog, found along the coast and ranges of New South Wales, and south-east Queensland, the Bleating Tree Frog is found in most available habitats.

Common around human habitation this frog is also often found around swamps and lagoons. This frog breeds in summer after thunderstorms and general rains.

The call is a distinctive long wavering bleat. Large numbers of this frog calling can produce a deafening noise.

74 Bleating Tree Frog *Litoria dentata* female (Ourimbah, NSW).

Bleating Tree Frog *Litoria dentata*

75 Eastern Dwarf Tree Frog *Litoria fallax* (Wyong, NSW).

EASTERN DWARF TREE FROG
Litoria fallax (Peters, 1880)
Fig. 75

Closely related to the Northern Dwarf Tree Frog *Litoria bicolor*, the Eastern Dwarf Tree Frog reaches 2 to 3 cm in length. It is found along the east coast of New South Wales and Queensland. The almost identical Northern Dwarf Tree Frog is found in tropical parts of northern Australia.

The Eastern Dwarf Tree Frog ranges from green to brown in colour. It is common around dams, lagoons, swamps and slow-moving creeks, where it is usually found during the day amongst vegetation, or foraging for small insects at night. This frog seems to survive well around waterways that are infested with mosquito fish *Gambusia* spp. which tend to eliminate most other frogs.

It breeds during the warmer months. Small clumps of eggs are laid in a dispersed manner amongst aquatic litter and submerged plants. The distinctive tadpoles have unusually prominent and well-pigmented tails. These attain some 5.5 cm before metamorphasing.

The call is a very loud 'wr-eeeek, wr-eeeek, wr-eeeek'.

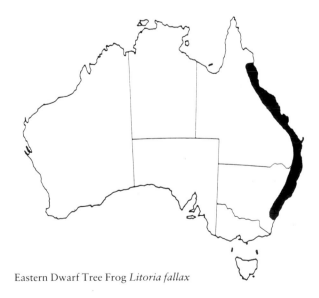

Eastern Dwarf Tree Frog *Litoria fallax*

76   Freycinet's Frog *Litoria freycineti* (West Head, NSW).

77   Broad-palmed Frog *Litoria latopalmata* (Dubbo, NSW).

78   Lesueur's Frog *Litoria lesueurii* (Ourimbah, NSW).

FREYCINET'S FROG
Litoria freycineti Tshudi, 1838
Fig. 76

Freycinet's Frog attains 5 cm in length, with females being by far the larger sex. This frog is found along the coast of New South Wales and south-east Queensland, in a variety of habitats.

A jumpy ground-dwelling species, it is usually found active on wet spring, summer and autumn nights foraging for insects. It breeds around small semi-permanent streams.

The call is a duck-like quacking at accelerating speed.

Freycinet's Frog *Litoria freycineti*

BROAD-PALMED FROG
Litoria latopalmata Gunther, 1867
Fig. 77

This 4-cm frog is variable in dorsal colour ranging from browns and greys to a fawnish colour. It is possible that more than one frog species are identified here. Found throughout the eastern third of Australia, this species occupies most habitats.

A ground-dwelling frog, it is usually found under cover such as rocks, logs, debris, etc. during the day.

It breeds during wet weather in summer around various types of watercourses, ranging from man-made to flood-plains of fast-flowing rivers.

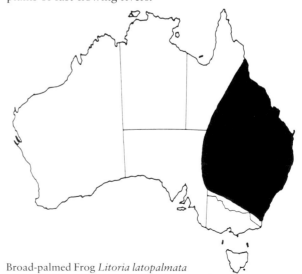

Broad-palmed Frog *Litoria latopalmata*

LESUEUR'S FROG
Litoria lesueuri (Dumeril and Bibron, 1841)
Fig. 78

Lesueur's Frog attains 7 cm in size but more than one frog species is probably included here. This frog is found along the coast and near ranges of eastern Australia from Victoria to the bottom of Cape York.

Found in a wide range of habitats this frog has similar habits to the Broad-palmed Frog *Litoria latopalmata*. This species covers great distances within a relatively short time when foraging. It is commonly encountered moving on wet nights when crossing roads.

In captivity it feeds on cockroaches and small beetles, and has been known to take smaller frogs. This species breeds in all but the coldest months of the year. The call is a soft purring trill repeated for a few seconds.

79 *Litoria* sp., closely related to *Litoria lesueurii* (Milla Milla, QLD).

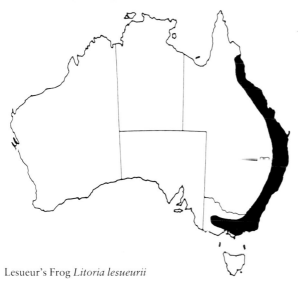

Lesueur's Frog *Litoria lesueurii*

ROCKET FROG
Litoria nasuta (Gray, 1842)
Fig. 80

The Rocket Frog *Litoria nasuta* is similar in appearance to *Litoria freycineti*. This 5-cm frog is found around the coastal and near coastal regions of north and east Australia.

This ground-dweller is usually found on wet nights

80 Rocket Frog *Litoria nasuta* (Kunnanurra, WA).

in summer foraging adjacent to the swamps and semipermanent creeks that it breeds in. It is a summer breeder, and the female is the larger sex.

The call of this frog is apparently the same as that of the Freycinet's Frog *Litoria freycineti*.

PERON'S TREE FROG
Litoria peronii (Tschudi, 1838)
Fig. 81

Found throughout New South Wales and adjacent areas, this frog attains 5 cm. The colour of this frog will change depending on the light exposure, temperature, moisture and other factors. Green flecks may be present or absent, and the dorsal colour will range from a dark grey to an almost white colour.

The Peron's Tree Frog is found in all types of habitat, often a long distance from water. This frog is often found under cover during the day, in places such as hollow trees, under bark, and sometimes around human habitations including water systems.

Rocket Frog *Litoria nasuta*

81   Peron's Tree Frog *Litoria peronii* female (Ourimbah, NSW).

82   *Litoria* sp., closely related to *Litoria peronii* (Milla Milla, QLD).

83   Leaf Green Tree Frog *Litoria phyllochroa* (West Head, NSW).

Specimens are often found moving about on wet nights. Although often occurring in association with large rivers and similar large bodies of water, this species breeds in the semi-permanent water bodies that form after rain in these areas, in summer. The tadpoles are an irridescent green colour with a high crested tail.

The call is a very distinctive chuckle consisting of between fifteen and twenty rapid descending notes.

Peron's Tree Frog *Litoria peronii*

LEAF-GREEN TREE FROG
Litoria phyllochroa (Gunther, 1863)
Fig. 83

A 4-cm frog, individuals of this species may change dorsal colour from dark or light green, to a purplish-brown colour.

This species occurs along the coast and nearby ranges of New South Wales and adjoining parts of Queensland and Victoria. A few almost identical, and closely related, species also occur in areas where this frog is found.

When inactive the Leaf-green Tree Frog is usually found in vegetation bordering creeks and waterholes during the warmer months. In winter this species is often found hibernating under rocks and logs in the vicinity of water. It is common around creeks in built-up parts of Sydney and Brisbane.

It breeds in spring and summer around small creeks and nearby waterholes, swamps, etc., and is commonly found

Leaf Green Tree Frog *Litoria phyllochroa*

active during the day in the breeding season. The female is the larger sex. The partially transparent, greyish-coloured tadpoles take about three months to turn into froglets.

The call sounds like a loud whirring 'wrk-wrk-wrk'.

WARTY GREEN AND GOLDEN BELL FROG
Litoria raniformis (Keferstein, 1867)
Fig. 84

The Warty Green and Golden Bell Frog is essentially a slightly larger and more robust form of the Green and Golden Bell Frog *Litoria aurea*. The Warty Green and Golden Bell Frog can be distinguished by its warty appearance and pale-green mid-dorsal stripe. Large females may attain up to 10 cm in length. Males tend to be substantially smaller. Tadpoles of this species are large and very fast-moving.

It is found throughout Tasmania, Victoria and adjacent cold parts of South Australia and New South Wales. Similar species occur in the New England region of New South Wales and south-west Western Australia.

This species is most commonly found around ponds in creeks, swamps, farm dams and similar bodies of water. It is usually seen by day resting in bullrushes or thick vegetation at the water's edge. The Warty Green and Golden Bell Frog commonly basks in the sun during the day. At night specimens don't tend to wander far and for this reason are very rarely seen crossing roads on wet nights, even in areas where this species is common.

The mating call is several short grunts followed by a long deep sonorous drone. When picked up, this species may emit a loud scream.

Warty Green and Golden Bell
Litoria raniformis

RED-EYED TREE FROG
Litoria rothi (De Vis, 1884)
Figs 85, 86

Similar in many respects to the Peron's Tree Frog *Litoria peronii*, this species is readily distinguished by the rusty-red colour of part of the eye. This 5-cm frog is found throughout non-arid tropical Australia.

Red-eyed Tree Frogs will commonly live around human settlements, in water tanks, drainage systems, etc. Usually found near large watercourses in the dry season, this species breeds in the many semi-permanent swamps that form during the wet season.

84. Warty Green and Golden Bell Frog *Litoria raniformis*, two males from Bundoora, Vic. Like some other species, specimens of this species can change in base colour from green to brown within ten minutes depending on environmental conditions.

85 Red-eyed Tree Frog *Litoria rothi* (Kunnanurra, WA).

86 Red-eyed Tree Frog *Litoria rothi* (Charters Towers, QLD).

87   Desert Tree Frog *Litoria rubella* (Moonie, QLD).

88   Splendid Tree Frog *Litoria splendida* (Lake Argyle, WA).

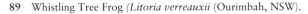
89   Whistling Tree Frog *(Litoria verreauxii* (Ourimbah, NSW).

The call is a loud chuckle consisting of about ten slightly drawn-out descending notes.

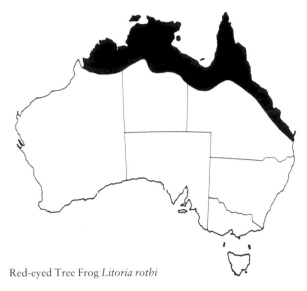

Red-eyed Tree Frog *Litoria rothi*

DESERT TREE FROG
Litoria rubella (Gray, 1842)
Fig. 87

This 3.5-cm frog varies in colour from grey to brown or fawn dorsally, and is similar in many respects to the Bleating Tree Frog *Litoria dentata*. It is found throughout mainland Australia, except for the far south.

This species occurs in all types of habitat although it always appears to be strongly associated with permanent water bodies, man-made dams or human habitation.

A summer breeder, this species is most commonly found active after summer rains, either breeding or feeding on small insects.

The call is a loud, high-pitched, distinctly pulsed note.

Desert Tree Frog *Litoria rubella*

SPLENDID TREE FROG
Litoria splendida Tyler, Davis and Martin, 1977
Fig. 88

Closely related to the Green Tree Frog *Litoria caerulea*, the Splendid Tree Frog *Litoria splendida* is readily distinguished by its distinctive yellowish or orange spots. This frog also has considerably enlarged parotid glands beneath the skin on the head.

The Splendid Tree Frog appears to be restricted to the Kimberley Ranges in Western Australia. In the dry season it is often found hiding in groups deep in rock crevices. In the wet season it usually breeds in nearby semi-permanent creeks and swamps. It will move into small water tanks and drainage systems if built in its area of occupation.

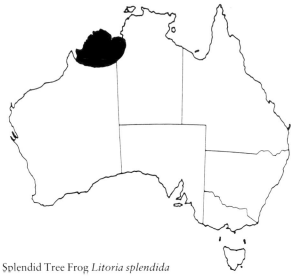

Splendid Tree Frog *Litoria splendida*

WHISTLING TREE FROG
Litoria verreauxii (Dumeril, 1853)
Figs 89, 90

This 3-cm frog is highly variable in dorsal colour. Although usually having some markings, this frog's base colour ranges from fawn, brown or russett in colour.

The Whistling Tree Frog is found throughout the coast, ranges and nearby areas of New South Wales, southeastern Queensland and eastern Victoria.

A common species, it can sometimes be found in massive numbers, particularly when breeding around manmade dams and swamps. It will, however, be found in most types of habitat and breeds in most types of permanent watercourse. Although most specimens breed in the warmer months, males will call all year. The call is a loud distinctive ascending whirring sound, and it is not uncommon for males to call during the day.

Whistling Tree Frog *Litoria verreauxii*

90 Whistling Tree Frog *Litoria verreauxii*, newly metamorphased young (Ourimbah, NSW).

91 Frog, species unknown (Mount Garnett, QLD).

92 Frog, species unknown (Wentworth Falls, NSW).

CLASS REPTILIA

Order Crocodilia

Family Crocodylidae

The order crocodilia contains crocodiles (Crocodylidae), alligators and caimans (Alligatonidae) and the gavial (Gavialidae), more than twenty species in all. All are found in tropical regions. Different types are distinguished by head morphology, as all are essentially similar in appearance.

Modern crocodiles have not changed in form significantly since the age of dinosaurs. All grow fairly large and are amphibious, usually living in larger watercourses where they occur. All are active predators, killing their prey by biting it, then rolling with it in water until it drowns, gets ripped apart or whatever. Adult crocodiles have no natural predators (except humans).

Crocodiles will bask during the day, but are most active at night. Despite their apparently sluggish nature, the crocodiles can move over land or in water at high speed when necessary.

Adults construct nests for their eggs, and parental care by adults of nest and young is well known.

Two species of crocodile are found in northern Australia: the Freshwater Crocodile and the Estuarine or Saltwater Crocodile.

93 Freshwater Crocodile *Crocodylus johnstoni* (Cairns, QLD).

FRESHWATER CROCODILE
Crocodylus johnstoni Krefft, 1873
Figs 93, 94

Found in permanent freshwater streams, rivers and billabongs in tropical Australia, from Cape York to the Kimberleys, this species averages 1.2 m but rarely exceeds 3 m. Although this species is usually only found in numbers in freshwater watercourses, there is at least one estuarine river habitat in the Northern Territory where this species lives and breeds in substantial numbers.

The Freshwater Crocodile is distinguished from the dangerous Saltwater Crocodile by its relatively long narrow snout. Although a few people have been attacked by large specimens of the Freshwater Crocodile, it is not generally regarded as being a threat to humans.

Its diet consists principally of fish, crustaceans, tortoises and sometimes small land-dwelling vertebrates.

Around August–September adult females lay about twenty eggs high up on sandbanks, often in floodplains. The 6-cm-long eggs are deposited in a dug hole about 45 cm deep and 22 cm in diameter. These eggs hatch around November, usually before the first floods of the wet season. The 25-cm hatchlings move into the adjacent water and appear to remain with the waiting female for an unknown period, before hiding among Pandanus roots and other bankside vegetation.

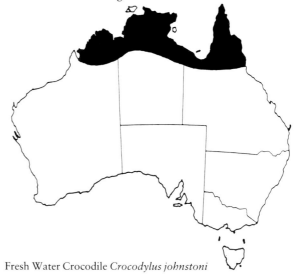

Fresh Water Crocodile *Crocodylus johnstoni*

SALTWATER OR ESTUARINE CROCODILE
Crocodylus porosus Schneider, 1801
Figs 95, 96

This dangerous species may exceed 7 m in length, although 4 m is a more common maximum length. It occurs throughout coastal North Australia and Queensland, and throughout South-east Asia.

94 Freshwater Crocodile *Crocodylus johnstoni*, head (Cairns, QLD).

Earlier this century crocodile-skin hunters almost exterminated this species. Populations are still recovering. Although populations remain severely depleted in coastal Queensland, the Saltwater Crocodile is now fairly common in northern Australia and it is in this area that most attacks on humans occur.

It is usually found in saltwater environments (hence its name), but will live in adjacent freshwater habitats. Some specimens are found many kilometres from coastal saltwater habitats. Other specimens are sometimes found crossing large bodies of saltwater such as the Timor Sea.

The diet of this species is varied but recent research points to a preference for red meat.

About fifty eggs are laid in the wet season in nests high up on river banks. The 8-cm-long eggs are laid in a nest of leaf mould some 0.45 m deep and 2 m in diameter. The female will often attempt to guard the nest against predators and to repair any damage done by them However, foxes, pigs and goannas commonly eat the eggs before they hatch, while other nests might be flooded out. Therefore most eggs laid during a season do not appear to hatch.

When the young hatch, they will yelp and be dug out of the nest by the female, usually waiting, who will then accompany the young in the water for about two days, before they disperse. When the young leave the nest they sometimes still have their yolk sacs attached. These either fall off, or are scratched off by the hind feet shortly after. Large numbers of young crocodiles often become the prey of Water Pythons *Bothrochilus fuscus*, birds, fish and pigs.

Specimens that wander into areas of human habitation are usually males.

In captivity Saltwater Crocodiles live for many years, and, relative to their body size, eat very little, due to their very slow metabolic rate. Hatchlings grow at the rate of about 30 cm a year, and then the rate slows down.

95   Saltwater Crocodile *Crocodylus porosus* (Cairns, QLD).

Saltwater Crocodile *Crocodylus porosus*

Order Testudines (Turtles and Tortoises)

Suborder Cryptodira (Sea Turtles)

Family Cheloniidae (Most Sea Turtles)

All but one large sea turtle, the Leatherback *Dermochelys coriacea*, belong to this family. All species generally only live in the tropics but some specimens do stray into temperate waters. Five species are known from Australian waters. All are characterised by hard shells, paddle-like flippers, and bony plates covering the shell.

Nesting occurs on a number of Australian beaches. The eggs are laid every few years (several times in one season), and the eggs take two to three months to hatch. The young, which usually have a high mortality rate, typically leave the nest after hatching at night. Most or all hatchlings leave a given nest at the same time, taking only a

96   Saltwater Crocodile *Crocodylus porosus* (Cairns, QLD).

matter of minutes to go from the nest to sea. The mortality rate is high, with crabs, birds, sharks all taking large numbers of young turtles. Adult specimens have few natural enemies. Sea turtles have little if any reason to come on to land except to nest. Mating turtles have been observed in shallow waters of tropical islands.

Intensive hunting by man of turtles and eggs has seriously threatened or endangered most species.

LOGGERHEAD TURTLE
Caretta caretta (Linnaeus, 1758)
Figs 97–99

The Loggerhead Turtle is found throughout tropical oceans of the world. In Australia it is regularly seen along the coasts of Queensland, the Northern Territory and the top half of Western Australia, but is most commonly seen along the Queensland coast. Specimens do occasionally stray further south, with specimens from Sydney known. This species attains a shell length of 1.5 m, and is distinguished by its colour and number of shields on the shell. The Loggerhead does not have a distinct beak, although many adults do have a very large head (hence the name 'loggerhead').

This species is mainly carnivorous, and the diet includes fish, etc, which are usually caught in deeper waters. Rookeries occur throughout northern Australia. Egg-laying occurs in the period October to late February, with eggs hatching about eight weeks later. The average number of eggs in a clutch is 120. The young measure about 3 to 5 cm (shell length).

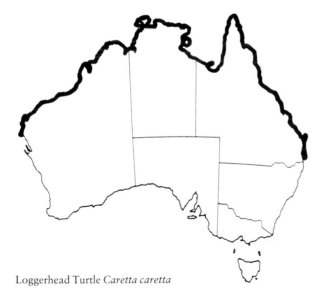

Loggerhead Turtle *Caretta caretta*

FLATBACK TURTLE
Chelonia depressa Garman, 1880
Fig. 100

Known only from the northern Australasian region, this species occurs along the coasts of the northern third of Australia, where it is fairly common in shallow waters. It is distinguishable by its flattened shell, and a thin fleshy skin covering the shell in adults. Adults attain a shell length of 1.2 m. Hatchling Flatback Turtles are nearly twice the size of those of Green Turtles or Loggerheads, being about 6 cm (shell length).

97   Loggerhead Turtle *Caretta caretta*, sub-adult (Hervey Bay, QLD).

98   Loggerhead Turtle *Caretta caretta*, hatchling (Mon Repos, QLD).

99   Loggerhead Turtles *Caretta caretta*, hatchlings moving down beach (Mon Repos, QLD).

The diet of this species is mainly carnivorous.

Egg-laying occurs between October and January, with the eggs taking about eight weeks to hatch. The female lays about sixty eggs to a clutch. Although this species nests on beaches throughout tropical Australia, most rookeries are located west of Cape York Peninsula.

100 Flatback Turtle *Chelonia depressa*, hatchling (Mon Repos, QLD).

Flatback Turtle *Chelonia depressa*

GREEN TURTLE
Chelonia mydas (Linnaeus, 1758)
Figs 101, 102

Found worldwide in tropical seas, the Green Turtle is common along the tropical coasts of the northern half of Australia. It attains more than a metre in shell length. It is probably the best known sea turtle, and is the species from which turtle soup is made.

An omnivorous species, the Green Turtle nests in spring

101 Green Turtle *Chelonia mydas* (Hervey Bay, QLD).

and summer, on various beaches throughout northern Australia. It lays about 110 eggs, which hatch about eight weeks later. Hatchlings measure 3–4 cm (shell length).

Green Turtle *Chelonia mydas*

102 Green Turtle *Chelonia mydas*, head (Hervey Bay, QLD).

103 Hawksbill Turtle *Eretmochelys imbricata* (Moreton Bay, QLD).

HAWKSBILL TURTLE
Eretmochelys imbricata (Linnaeus, 1766)
Figs 103–105

The Hawksbill Turtle is found in warmer seas worldwide. Named 'Hawksbill' because of the distinctive 'beak' present on the head, this species is very common in Australia's tropical waters. It grows to about a metre (shell length). This is the species from which tortoiseshell, used in the making of various expensive artefacts, is derived.

This turtle is largely carnivorous and is often found around coral reefs where it feeds on a variety of fish, including small sharks.

In Australia this species is only known to nest around the Gulf of Carpentaria and Torres Strait. About fifty eggs are laid per clutch.

104 Hawksbill Turtle *Eretmochelys imbricata*, head (Moreton Bay, QLD).

105 Hawksbill Turtle *Eretmochelys imbricata*, swimming (Green Island, QLD).

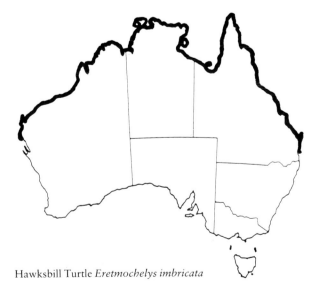

Hawksbill Turtle *Eretmochelys imbricata*

106   Broad-shelled Tortoise *Chelodina expansa* (Burnett River, QLD).

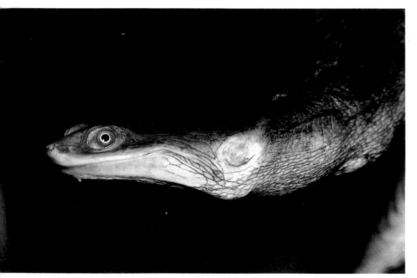

107   Broad-shelled Tortoise *Chelodina expansa*, head (Burnett River, QLD).

108   Long-necked Tortoise *Chelodina longicollis*, male, 'narrow-shelled' form found east of the Great Dividing Range (Lane Cove, NSW).

Suborder Pleurodira (Sideneck Tortoises)

Family Chelidae (Freshwater Tortoises)

Called terrapins in other countries, this particular family has species in Australia, New Guinea and South America. At least sixteen species are currently recognised within Australia.

No Australian species is threatened in any way, except for one species near Perth, Western Australia, in which less than a thousand individuals remain. Current policies of the West Australian Government, effectively prohibiting any captive breeding programmes, are ensuring the rapid extinction of that species, the Western Short-necked Tortoise *Pseudemydura umbrina*.

Freshwater tortoises have well-developed webbed feet, and can usually withdraw their heads into their shells. Their shells are covered by well-defined horny plates. Despite the high visibility of most species, many remain poorly known and undescribed, even on the heavily populated east coast of Australia.

Like sea turtles, all species lay eggs, but the nature of the eggs laid varies, as do nesting habits between species, and even within geographical races of the same species.

BROAD-SHELLED TORTOISE
Chelodina expansa Gray, 1857
Figs 106, 107

This large tortoise attains a 50-cm shell length. It is typified by its extremely long neck and flattened head, and is the largest long-necked species.

This tortoise is essentially only a river species, although around Moonie, in Queensland, specimens appear to make large overland journeys when it rains. It is found in coastal south-eastern Queensland, including Fraser Island, as well as in the Murray–Darling river system. Specimens from the Murray–Darling system have very flattened shells, while those from coastal Queensland (usually the Burnett River system) have a deeper shell (pictured), as do older specimens from elsewhere.

Broad-shelled Tortoise *Chelodina expansa*

This species strikes at its food in a manner not unlike that of a snake striking its prey from a coiled position. Captive specimens are known to have fed on small chickens and mice, as well as more typical foods.

The large hard-shelled eggs, when laid in late summer and autumn, appear to take nearly twelve months to hatch. Hatchlings measure about 3.5 cm (shell length), but can more than double this size within twelve months if well fed.

LONG-NECKED TORTOISE
Chelodina longicollis (Shaw, 1794)
Figs 108–112

This tortoise can attain 30 cm in shell length, although most adult specimens average 15 to 20 cm. The species is found throughout south-eastern and eastern Australia, and it is the species most commonly kept as a pet in eastern Australia.

Its dorsal shell (carapace) colour can vary from all black through all shades of brown to almost white, with well-defined darker lines around the scutes. Specimens

109 Long-necked Tortoise *Chelodina longicollis*, male, 'broad-shelled' form found west of the Great Dividing Range (Bingarra, NSW).

Eastern Long-necked Tortoise
Chelodina longicollis

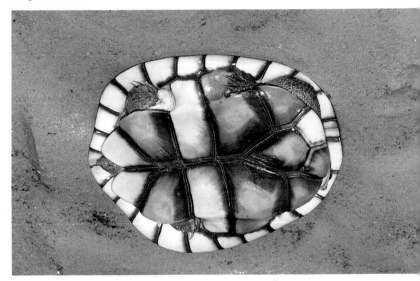

110 Long-necked Tortoise *Chelodina longicollis*, male, ventral surface (Bingarra, NSW).

111 Long-necked Tortoise *Chelodina longicollis*, juvenile, captive bred by author from 'mixed' parents.

112 Long-necked tortoise *Chelodina longicollis*, juvenile, captive bred by author from 'mixed' parents. Ventral surface.

from coastal areas tend to have darker narrower shells than those from the Murray–Darling river system. It is found in most water systems and major freshwater bodies within its range, including small creeks. During summer major overland migrations may occur.

Although an opportunistic feeder, this species shows a strong preference for frogs and snails.

Egg-laying occurs in spring and early summer, when the female tortoise deposits four to twelve soft-shelled eggs in an excavated hole adjacent to the water where it lives. These usually hatch in autumn but sometimes in mid winter (100–200 days later).

The young tortoises are usually brightly coloured with black shell and orangish markings which fade with age.

OBLONG TORTOISE
Chelodina oblonga Gray, 1841
Fig. 113

The Oblong Tortoise attains a shell length of up to 31 cm and is usually found in permanent waterways of south-western Western Australia. Carapace colour ranges from grey to olive-brown in colour.

Of all the long-necked species the Oblong Tortoise has the longest neck in proportion to its shell and body. The head and neck may extend up to 90 per cent of the carapace (shell) length. The head and neck are far too large and thick to receive any significant protection when withdrawal into the shell is attempted.

Regular seasonal overland migrations do occur, and in some areas near Perth, Western Australia, road signs warn motorists of tortoises crossing the roads.

When walking overland, this tortoise moves fairly rapidly. The head and neck are held straight in front of the carapace, giving the appearance that the tortoise is determined to get somewhere.

Egg-laying is in October–November, and the eggs take up to 200 days to hatch. The carapace of juveniles is slightly expanded at the rear, but this broadness reduces with age.

Oblong Tortoise *Chelodina oblonga*

113 Oblong Tortoise *Chelodina oblonga* (locality unknown).

NORTHERN LONG-NECKED TORTOISE
Chelodina rugosa Ogilby, 1890
Figs 114, 115

The Northern Long-necked Tortoise, with a 35-cm maximum shell length, is found in most major watercourses in far northern Australia.

The large head and neck of this species is afforded little protection by the shell when retracted. Some specimens of this species have very prominent barbells on the chin, more so than in other tortoises of the genus *Chelodina*.

Egg-laying is believed to be at the end of the wet season between March and May, with the eggs presumably hatching at the commencement of the next wet season

114 Northern Long-necked Tortoise *Chelodina rugosa* (Alligator River, NT).

Northern Long-necked Tortoise
Chelodina rugosa

around December–January. An average of twelve eggs are laid at a time, and this species is known to produce multiple clutches within a short period of time. Eggs measure 3.5 cm by 2.5 cm.

Wild specimens are believed to feed on crustaceans, fish and other things. Captive specimens thrive on various types of meat.

NORTHERN SNAPPING TORTOISE
Elseya dentata (Gray, 1863)
Figs 116, 117

More than one species may be recognised here. This tortoise is found in most river systems north of the Tropic of Capricorn. There appears to be a strong geographical variation in appearance and size of the specimens. Like other species of tortoise in the genus *Elseya*, this species usually has a yellow plastron (under shell), but there is a strong tendency to melanism in older specimens. The largest specimens attain more than 45 cm in size (females), but the average adult size is about 30 cm (shell lengths).

Adults of this species often have disproportionately large heads (macrocephaly), and are capable of giving a serious bite.

In the wild this species feeds on crustaceans, fish and possibly various tropical fruits that fall into the rivers that they inhabit.

This tortoise lays the largest eggs of any Australian tortoise species, measuring 5.5 cm in length. These hard-shelled eggs are laid at the end if the northern wet season and incubate and hatch in the northern dry season.

Northern Snapping Tortoise *Elseya dentata*

EASTERN SNAPPING TORTOISE
Elseya latisternum Gray, 1867
Figs 118–120, 126

This tortoise has a 20-cm shell length and is found throughout coastal Queensland and northern New South Wales. It is found in various types of watercourse, but is most common in medium to large and fast-flowing rivers, where it is sometimes found in relatively large numbers.

The Eastern Snapping Tortoise feeds on fish and other animals found within its habitat.

Nesting occurs in spring when females lay eggs in holes dug high up on riverbanks. The dozen or so eggs hatch about two to three months later.

115 Northern Long-necked Tortoise *Chelodina rugosa*, ventral surface (Alligator River, NT).

116 Northern Snapping Tortoise *Elseya dentata*, female (locality unknown, but probably QLD).

117 Northern Snapping Tortoise *Elseya dentata*, female, head (locality unknown, but probably QLD).

Eastern Snapping Tortoise *Elseya latisternum*

118   Eastern Snapping Tortoise *Elseya latisternum*, female (Burnett River, QLD).

119   Eastern Snapping Tortoise *Elseya latisternum*, male, ventral surface showing melanism (typical of adult *Elseya*) (Burnett River, QLD).
120   Eastern Snapping Tortoise *Elseya latisternum*, hatchling, ventral surface. Captive bred; parents from Burnett River, QLD.

KREFFT'S TORTOISE
Emydura krefftii (Gray, 1871)
Figs 8, *121–123*, *126*

This tortoise has a 25-cm shell length and is found throughout Queensland. In the past this species was used by Queensland Aborigines as a major source of food. Some specimens have distinct yellow markings on the head, while on others these are absent. It is essentially a river-dwelling species.

On Fraser Island, Queensland, this species is extremely common in sandy lagoons. Specimens here do not attain the same adult size as mainland tortoises. Research shows this is due to changed environmental conditions on the island and the restricted gene pool adapting to local conditions without outside interference.

The Krefft's Tortoise lays eggs in spring, which subsequently hatch in summer. The 2.5-cm shell length hatchlings emerge some eighty days after laying.

121   Krefft's Tortoise *Emydura krefftii*, female (Burdekin River, QLD).

Krefft's Tortoise *Emydura kreffti*

122 Krefft's Tortoise *Emydura krefftii*, hatchling, captive bred. Parents from Burnett River, QLD.

MACQUARIE TORTOISE
Emydura macquarii (Gray, 1830)
Fig. 124

The Macquarie Tortoise attains 30 cm (shell length) and is restricted to the Murray–Darling river system. This species prefers to inhabit main course waterways.

Its entire range is within the Australian 'food bowl' areas, and many specimens are caught by fishermen or in nets. Predation of nests by foxes, cats, pigs and wild dogs appears to be taking a major toll on numbers, too, which are in sharp decline throughout its range.

About ten elongated eggs are laid in spring and summer in a hole dug on ground adjacent to the river. These hatch about eighty days later.

The Macquarie Tortoise feeds principally on fishes, molluscs and crustaceans.

Like most Australian short-necked tortoises, the male has a much bigger and longer tail than the female.

123 Krefft's Tortoise *Emydura krefftii*, hatchling, ventral surface. Note missing hind limb. Captive bred specimen. Parents from Burnett River, Qld.
124 Macquarie Tortoise *Emydura macquarii*, male (locality unknown).

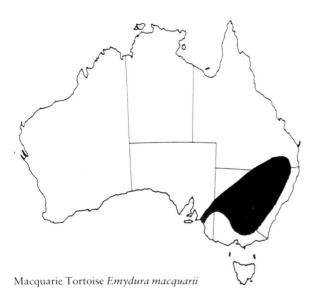

Macquarie Tortoise *Emydura macquarii*

125   Eastern Short-necked Tortoise *Emydura signata* (locality unknown).

Eastern Short-necked Tortoise *Emydura signata*

126   Krefft's Tortoise and Eastern Snapping Tortoise captive bred hatchlings.

EASTERN SHORT-NECKED TORTOISE
Emydura signata Ahl, 1932
Fig. 125

This small, 20-cm shell length tortoise is similar in most respects to the Macquarie Tortoise *Emydura macquarii*. It is found in rivers of the north coast of New South Wales and the south Queensland coast.

It is possible that more than one species of tortoise is being grouped here. Recent attempts at reclassifying the Eastern Short-necked Tortoise *Emydura signata* have focused on different eye characteristics and colour as well as shell morphology, etc. It seems that almost every major river system in this area has its own distinctive form of *Emydura signata*.

These tortoises are spring breeders, but their biology is little known.

Order Squamata (Snakes and Lizards)

Suborder Sauria (Lizards)

Family Agamidae (Dragon Lizards)

This family is found throughout the old world. More than fifty species occur in Australia, mostly in drier areas.

Dragons are characterised by their juxtaposed scalation, often including distinctive spines. They range in adult size from 10 cm to a metre.

Many species possess pre-anal and femoral pores, which are tiny holes located in front of the anus and on the ventral surface of the hind legs. The function of these pores is believed to be sexual.

All are of the typical lizard form, generally diurnal and egg-layers. Species do not routinely 'throw' or loose their tails, although some (notably Water Dragons *Physignathus* spp.) can regenerate a lost tail.

MOUNTAIN DRAGON
Amphibolurus diemensis (Gray, 1841)
Figs 127, 128

This 18-cm lizard is found along the coast and ranges of south-eastern Australia, south of Wyong (New South Wales) and scattered highland areas in the New England region (New South Wales). Victorian specimens seem to grow larger than their New South Wales counterparts. Colour varies widely between individuals even within a given locality. The two sexes also have different patterns.

Commonly confused with other types of dragon, the

Mountain Dragon may be distinguished by the blue interior of the mouth.

The Mountain Dragon is most common in low scrub, heaths and similar habitats. In cold areas this species may be found in a nearly frozen state in winter. It is insectivorous.

The female lays about two to eight eggs in a sheltered spot in early summer. Hatchlings measure about 3 cm and are generally dull in colour.

Mountain Dragon *Amphibolorus diemensis*

JACKY
Amphibolurus muricatus (White, ex Shaw, 1790)
Figs 11, 129–131

The Jacky averages 30 cm and is found along the coast and ranges of south-eastern Australia. When cornered it will puff up and open its mouth wide open, exposing a bright orange colour.

Males and females have different colour patterns (illustrated) and colour darkness responds directly to temperature at the given time (the hotter the lizard, the lighter the colour).

Found in woodlands, sclerophyll forests and heaths, this species appears to become particularly abundant in areas recently affected by bushfires, often moving in from unburned, less-preferred habitats.

Jacky *Amphibolorus muricatus*

127 Mountain Dragon *Amphibolurus diemensis* (Newnes, NSW).

128 Mountain Dragon *Amphibolurus diemensis*, hatchling (Wentworth Falls, NSW).

129 Jacky *Amphibolurus muricatus*, male (North Ryde, NSW).

When pursued in the bush it readily takes to a tree, which it will actively keep between itself and the aggressor.

An insect feeder, it lays from four to twelve eggs in late spring–early summer. Hatchlings measure about 7 cm.

130 Jacky *Amphibolurus muricatus*, female (North Ryde, NSW).

131 Jacky *Amphibolurus muricatus*, female, threat display (North Ryde, NSW).

132 Frill-necked Lizard *Chlamydosaurus kingii* (Kunnanurra, WA).

FRILL-NECKED LIZARD
Chlamydosaurus kingii Gray, 1825
Figs *132–134*

This 70-cm lizard, depicted on the two-cent coin, is possibly the best known in Australia. It is characterised by a large fold of skin around the neck, which the lizard erects by opening its mouth when alarmed. The frill is also believed to aid thermoregulation.

The Frill-necked Lizard is found throughout non-arid tropical Australia. Its ground colour varies considerably, with specimens from the north and north-west of Australia being brightest in colour. Queensland specimens tend to be a drab brown or grey in colour.

Living in woodland habitats, this lizard is also well known for running away on its two hind legs when disturbed sunning itself or foraging. This lizard does not hesitate to take to trees when under threat. If cornered it will also lash its tail, and as a last resort bite. The Frill-necked Lizard has large teeth and powerful jaws so a bite can be very painful.

The lizard is most active during the wet season, when most specimens are caught. This species generally does not do well in captivity.

It feeds mainly on small insects, smaller reptiles and

Frill-necked Lizard *Chlamydosaurus kingii*

133 Frill-necked Lizard *Chlamydosaurus kingii*, head (Kunnanurra, WA).

some vegetable material. Mating is in September with about twelve to sixteen eggs being laid in November. The eggs are about 28.5 mm long by 20 mm wide, weighing about 5.3 grams. The 12–14.5 cm hatchlings emerge in early February. They weigh from 3.4 to 5 grams.

TAWNY DRAGON
Ctenophorus decresii (Dumeril and Bibron, 1837)
Fig. 135

The Tawny Dragon attains about 21 cm in length, and is highly variable in coloration. Males and females have markedly different colorations and colour patterns. This lizard is found in parts of south-eastern South Australia and parts of north-western New South Wales, usually in rocky habitats.

It is mainly insectivorous and specimens are usually seen actively foraging in hot weather.

Tawny Dragon *Ctenophorus decresii*

CENTRAL NETTED DRAGON
Ctenophorus nuchalis (De Vis, 1884)
Fig. 136

This 20–25-cm lizard is common throughout arid parts of all mainland states except Victoria, especially in areas with sandy or loamy soils.

Active specimens are usually seen when perched on elevated cover, such as logs, rocks or vegetation. They flee when approached and take to a shallow burrow or other

Central Netted Dragon *Ctenophorus nuchalis*

134   Frill-necked Lizard *Chlamydosaurus kingii*, head, dull coastal Queensland form (Cardwell, QLD).

135   Tawny Dragon *Ctemphorus decresii*, female (Middleback Ranges, SA).

136   Central Netted Dragon *Ctenophorus nuchalis* (Kingoonya, SA).

cover. Large numbers of specimens are often found around human rubbish dumps, where they shelter under sheets of metal, cardboard and other rubbish.

This mainly insectivorous lizard is active in very high temperatures.

137 Painted Dragon *Ctenophorus pictus*, male (Andamooka, SA).

PAINTED DRAGON
Ctenophorus pictus (Peters, 1866)
Fig. 137

This species attains 25 cm and is found throughout arid parts of the south half of the continent. The Painted Dragon is swift-moving and usually lives in red soil country and sand dunes where it occurs. It occupies small burrows concealed within low vegetation such as saltbush.

Males will raise their dorsal crest when alarmed.

By day this insectivorous lizard forages in litter and open ground near its concealed burrow to which it will retreat if disturbed.

Breeding habits are little known.

GREEN DRAGON
Diporiphora superba Storr, 1974
Fig. 138

The Green Dragon grows to 36 cm and is restricted to the Kimberley Ranges of north-western Western Australia. The tail accounts for more than three-quarters of this lizard's length.

This lizard is perfectly camouflaged when moving about in low shrub vegetation, which is its preferred habitat. Quite often this lizard goes unnoticed, even when common, because often it won't move when approached while concealed in vegetation.

The Green Dragon feeds on insects and other soft-bodied animals.

Painted Dragon *Ctenophorus pictus*

Green Dragon *Diporiphora superba*

A 35-cm lizard, the Southern Angle-headed Dragon is found in rainforests from the central coast of New South Wales to south-eastern Queensland. A similar species occurs in north-eastern Queensland. The prominent spikes are always erect. This species is diurnal and mainly aboreal.

This species relies on camouflage to avoid detection, and is usually only found when crossing roads and bush tracks. It is generally not a very fast-moving species.

Little is known of its biology although it appears to be insectivorous.

GILBERT'S DRAGON
Lophognathus gilberti Gray, 1842
Fig. 141

This 40-cm lizard is common throughout the northern third of Australia. Although found in a wide range of habitats it is usually found in association with water courses. The colour varies from grey to brown or reddish.

A very agile species, it usually takes to a tree when disturbed. If disturbed in an open area, the Gilbert's Dragon will run on its two hind legs at very high speed. It feeds principally on insects.

Males have a more distinct pattern than females, and are usually the larger sex.

This lizard is an egg-layer, although nothing is known of its breeding biology.

138 Green Dragon *Diporiphora superba* (Kununurra, WA).

SOUTHERN ANGLE-HEADED DRAGON
Gonocephalus spinipes (Dumeril and Dumeril, 1851)
Fig. 140

139 *Diporiphora* sp. (Lake Argyle, WA).

140 Southern Angle-headed Dragon *Gonocephalus spinipes* (Mount Glorious, QLD).

Southern Angle-headed Dragon
Gonocephalus spinipes

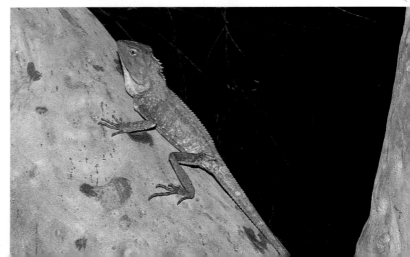

Gilbert's Dragon *Lophognathus gilberti*

141 Gilbert's Dragon *Lophognathus gilberti* (Arnhem Hwy, NT).

142 *Lophognathus* sp. (Kunnanurra, WA).

143 Agamid from Kunnanurra, WA.

GIPPSLAND WATER DRAGON
Physignathus howittii McCoy, 1878
Figs 144—146

A large 65-cm lizard, the male Gippsland Water Dragon is usually the larger sex and also by far the most brightly coloured.

This usually shy species is found from south and south-west of Sydney to eastern Victoria. It is usually found along fast-flowing rocky creeks and rivers throughout its range, although it may be found in association with various types of watercourse.

The Gippsland Water Dragon generally climbs trees or takes to water when frightened, being agile and fast-moving. When swimming this lizard holds its limbs against the sides of its body and moves in a snake-like manner. In certain tourist areas specimens lose their fear of humans and will take food from human hands.

It is omnivorous, feeding on most small enough vertebrates, insects and various types of fruits and berries.

About ten eggs are laid in summer, either under a rock or at the end of a specially dug burrow. They hatch some eighty days later.

Gippsland Water Dragon *Physignathus howittii*

EASTERN WATER DRAGON
Physignathus lesueurii (Gray, 1831)
Fig. 147

The Eastern Water Dragon usually measures 65 cm, although some males exceed 90 cm. It lacks the blue and green colours found in the Gippsland Water Dragon *Physinathus howittii*. Adult males instead have bright blood-red bellies.

The Eastern Water Dragon is found along the east coast from south of Sydney to Cape York. This adaptable lizard is found on all shores of Sydney Harbour and is common in heavily built-up areas. Its diet is mixed.

These lizards are known to inhabit rocky areas adjacent to beaches as well as living around mangrove swamps. A preferred habitat of this lizard is areas infested with the introduced weed Lantana, which afford it considerable protection. It is common for specimens to drop from a great height out of a tree and into a creek below when disturbed.

Fighting between males during the breeding season is common. Males appear to be territorial and all females within a male's territory are under his 'control'. Immature

144 Gippsland Water Dragon *Physignathus howittii*, male (Abercrombie, NSW).

145 Gippsland Water Dragon *Physignathus howittii*, male, head (Abercrombie, NSW).

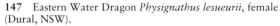

Eastern Water Dragon *Physignathus lesueurii*

146 Gippsland Water Dragon *Physignathus howittii*, juvenile (Abercrombie, NSW).

147 Eastern Water Dragon *Physignathus lesueurii*, female (Dural, NSW).

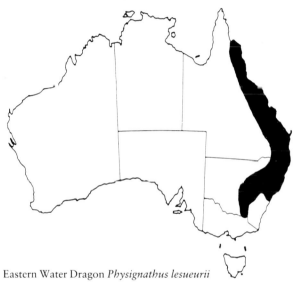

148 Bearded Dragon *Pogona barbatus* (St Clair, NSW).

149 Bearded Dragon *Pogona barbatus*, juvenile (St Clair, NSW).

specimens are apparently not affected by the territorial behaviour of adult males.

The ten or so eggs are laid under cover in well-concealed locations, in summer which hatch about eighty days later. Young specimens lack bright adult colours.

BEARDED DRAGON
Pogona barbatus (Cuvier, 1829)
*Figs 13, 14, **148, 149***

The Bearded Dragon averages 45 cm and is found throughout eastern Australia. Similar related species are found throughout Australia. It is commonly confused with the better-known Frill-necked Lizard *Chlamydosaurus kingii*. The Bearded Dragon *Pogona barbatus* is usually greyish in colour, although brick-red specimens do occur. Individuals of this species may change colour, depending on health, temperature and other factors. As a rule of thumb specimens become lighter when hotter, and darker when cooled.

The Bearded Dragon will puff up and raise its beard when startled, simultaneously exposing its bright-yellow mouthparts with mouth agape. Specimens are commonly found in all types of habitat, usually basking in an elevated spot such as a fencepost.

Diet consists of live arthropods, worms, etc., and some types of vegetation including flowers. Like many flower-eating lizards, this species seems to show a preference for yellow ones.

150 North-western Dwarf Bearded Dragon *Pogona mitchelli* (Shay Gap, WA).

About ten to twenty eggs are laid in spring. The eggs are laid in a hole about 18 cm deep in sandy soil. They hatch about seventy days later.

I observed a 'nest' constructed by this species in the middle of a popular walking-track/fire-trail, under the constant harassment and observation of hikers.

151 Interior Bearded Dragon *Pogona vitticeps* (Warren, NSW).

Bearded Dragon *Pogona barbatus*

NORTH-WESTERN DWARF BEARDED DRAGON
Pogona mitchelli (Badham, 1976)
Fig. 150

Found in the Pilbara and adjacent areas, this 30-cm lizard is essentially a dwarf-form of the Bearded Dragon *Pogona barbatus*.

It is usually found active in the mornings and late afternoon when it is not too hot. It feeds on insects, small lizards and flowers.

Little is known of the biology of this species, although it is presumed to be similar to that of the Bearded Dragon *Pogona barbatus*.

152 Interior Bearded Dragon *Pogona vitticeps* (Inland, Western Victoria).

INTERIOR BEARDED DRAGON
Pogona vitticeps Ahl, 1926
Figs 151, 152

This 45-cm lizard is very similar to the Bearded Dragon *Pogona barbatus* from which it can be distinguished by its more robust body and the configuration of spines on the

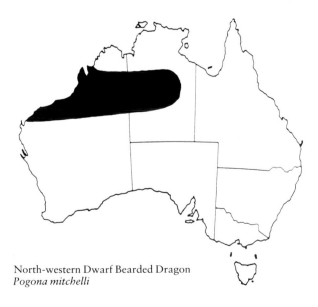

North-western Dwarf Bearded Dragon
Pogona mitchelli

Interior Bearded Dragon *Pogona vitticeps*

head and body. It often has a reddish tinge to its body colour, particularly around the eye.

This species can usually be seen basking on fenceposts along roads in areas where it occurs. Specimens are most visible in the months of September–October, when daytime temperatures seem to be most conducive to its activity.

It feeds on insects and flowers, and lays about twenty eggs in spring or summer.

EARLESS DRAGON
Tympanocryptis cephalus Gunther, 1867
Fig. 153

This small 10-cm lizard occurs in inland parts of Queensland, South Australia, the Northern Territory and much of Western Australia. Similar species are found in drier parts of all mainland states.

Earless Dragons are identifiable by the lack of an external ear opening. The ear is actually present below a layer of skin.

Earless Dragons are insectivorous and can commonly be found eating insects killed by cars on roads at dusk. Specimens have been found sleeping on roads at night in warmer weather. Often individuals will occupy burrows sited at the base of a low shrub or rock, to which they will retreat when alarmed. When foraging this lizard may stand on its hind limbs and tail (forming a tripod), with the forelimbs held against the side of the body in a bid to get a better view of the surrounding countryside. This position may be maintained for more than ten minutes.

Breeding biology is little known.

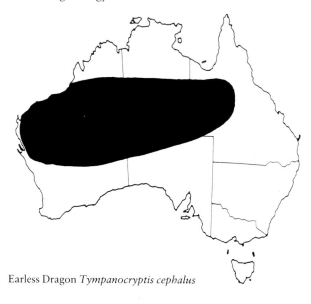

Earless Dragon *Tympanocryptis cephalus*

153   Earless Dragon *Tympanocryptis cephalus* (Hughenden, QLD).

Family Gekkonidae (Geckoes)

Found throughout the warmer parts of the world, more than sixty species of gecko are found in Australia. Geckoes are small lizards ranging in size from 6 to 30 cm. They are characterised by their juxtaposed scalation and, with a few spiny exceptions, most are soft-bodied. All are of the typical lizard form, generally nocturnal and egg-layers, usually laying two eggs at a time. Most species can lose and then subsequently regenerate their tails when attacked by predators. Many species can change colour in response to temperature, light intensity and other factors.

Many geckoes have modified toes with 'suckers' on them which enable them to climb most types of surface including glass. These suckers are formed by ultra-fine hairs on the lamellae (toe pads).

Most adult male geckoes can be identified by their enlarged anal region.

SPINY-TAILED GECKO
Diplodactylus ciliaris Boulenger, 1885
Figs 157–160

This gecko varies considerably in colour and morphology throughout its range. It is found throughout north-western and inland Australia. More than one species may be identified here.

The average size is 12 cm. In forest areas this is an arboreal species, while in deserts it inhabits spinifex grass, *Triodia* spp. It forages in open areas at night for insects.

These geckoes can exude a sticky substance from spines on their tails, which are actually pores between the small scales. This is a defensive mechanism and is employed when the gecko is under attack. Specimens will rarely shed their tails when threatened.

Spiny-tailed Gecko *Diplodactylus ciliaris*

FAT-TAILED DIPLODACTYLUS
Diplodactylus conspicillatus Lucas and Frost, 1897
Fig. 161

This gecko is common throughout most drier parts of Australia and attains 9 cm. The dorsal colour is highly variable, but the distinct body form of this gecko makes it easy enough to identify.

Occupying most habitats where it occurs, this species is a ground-dweller, sheltering under ground debris or in burrows under spinifex bushes during the day. By night it forages in the open for insects.

The distinctive fat tail is used as a food and water store during times when these are scarce. The tail is also used to plug its burrow when retired during the day.

When caught by a predator, this species will puff up its body in a bid to make it harder to swallow.

154   Agamid from Lake Argyle, WA.

Fat-tailed Diplodactylus
Diplodactylus conspicillatus

EASTERN SPINY-TAILED GECKO
Diplodactylus intermedius Ogilby, 1892
Fig. 162

This gecko grows to 10–12 cm and is found throughout semi-arid parts of southern Australia and the Macdonnell Ranges in the Northern Territory.

Most specimens are found during the day under tree bark, or in upright tree hollows. In areas without trees specimens may be caught under rocks.

This nocturnal lizard is able to exude a sticky substance from the spines in its tail when harassed.

155   Agamid from Lake Argyle, WA.

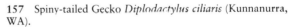

157   Spiny-tailed Gecko *Diplodactylus ciliaris* (Kunnanurra, WA).

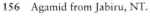

156   Agamid from Jabiru, NT.

158 Spiny-tailed Gecko *Diplodactylus ciliaris*, head (Kunnanurra, WA).

Eastern Spiny-tailed Gecko
Diplodactylus intermedius

159 Spiny-tailed Gecko *Diplodactylus ciliaris* Barkly Tableland, NT).

160 Spiny-tailed Gecko *Diplodactylus ciliaris* (Mount Isa, QLD).

161 Fat-tailed Diplodactylus *Diplodactylus conspicillatus* (Shay Gap, WA).

STEINDACHNER'S GECKO
Diplodactylus steindachneri Boulenger, 1885
Fig. 163

This 9-cm gecko is found throughout most of inland New South Wales and most parts of Queensland. The Steindachner's Gecko occupies all types of habitat except wet forests. This terrestrial species is usually found active in open spaces at night. By day it shelters in abandoned insect burrows, cracked mud in floodways, and elsewhere.

Its breeding biology is not known.

Steindachner's Gecko
Diplodactylus steindachneri

CROWNED GECKO
Diplodactylus stenodactylus Boulenger, 1896
Fig. 164

This 8-cm gecko is found in most drier parts of Australia, and tropical areas of the Northern Territory and Western Australia.

This terrestrial species occupies a variety of habitats, and may be found either during the day under cover or foraging at night after insects.

Crowned Gecko *Diplodactylus stenodactylus*

162   Eastern Spiny-tailed Gecko *Diplodactylus intermedius* (Bourke, NSW).

163   Steindachner's Gecko *Diplodactylus steindachneri* (Charleville, QLD).

164   Crowned Gecko *Diplodactylus stenodactylus* (Lake Argyle, WA).

165 Spinifex Striped Gecko *Diplodactylus taeniatus*, (Barkly Tableland, NT).

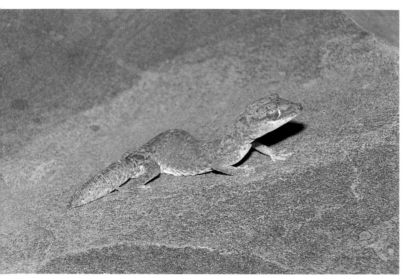

166 Tesselated Gecko *Diplodactylus tessellatus* (Charleville, QLD).

167 Stone Gecko *Diplodactylus vittatus* (Ourimbah, NSW).

SPINIFEX STRIPED GECKO
Diplodactylus taeniatus (Lonnberg and Andersson, 1913)
Fig. 165

Found throughout drier parts of north and north-western Australia, this small gecko only attains 7 cm, and is of very slender build.

It appears to be strictly associated with spinifex *Triodia* spp. and similar related grasses in which it lives. When within the spinifex bushes this lizard moves like a stick insect. Its tail is prehensile, and rarely shed. Most specimens are often seen crossing roads at night in spinifex country.

This lizard is capable of exuding an annoying sticky substance from its tail, and, when caught, it will open its mouth widely, displaying a bright orange interior.

Spinifex Striped Gecko *Diplodactylus taeniatus*

TESSELLATED GECKO
Diplodactylus tessellatus Gunther, 1875
Fig. 166

The colour of this gecko is highly variable, ranging from greys, browns and reds with various patterns dorsally, but it is always recognisable by its drab appearance. Its colour usually correlates with the ground colour where it occurs.

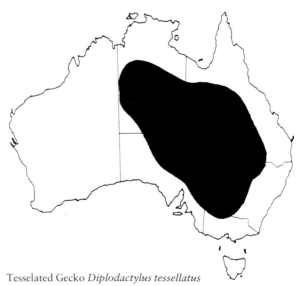

Tesselated Gecko *Diplodactylus tessellatus*

It attains 8 cm and is found throughout drier parts of the eastern half of Australia, including inland Victoria.

A terrestrial species, it is found during the day under various forms of cover, or soil cracks. By night it feeds on insects in open areas.

STONE GECKO
Diplodactylus vittatus Gray, 1832
Fig. 167

This 8-cm gecko is common throughout south-east Australia. Juvenile specimens are commonly black in colour.

A ground dweller, it is usually found during the day under cover such as rocks, logs, etc. In forest areas this species prefers exposed places such as large flat rock outcrops, in which they usually occupy rock on rock (as opposed to rock on dirt) positions. At night this species prefers to forage after insects in open spaces.

This species used to be common in the Rookwood cemetery in the heart of Sydney's western suburbs.

Stone Gecko *Diplodactylus vittatus*

NORTH-EASTERN SPINY-TAILED GECKO
Diplodactylus williamsi Kluge, 1963
Fig. 168

North-eastern Spiny-tailed Gecko
Diplodactylus williamsi

168   North-eastern Spiny-tailed Gecko *Diplodactylus williamsi* (Boggabri, NSW).

169   Spotted Dtella *Gehyra punctata* (Bungle Bungles, WA).

The North-eastern Spiny-tailed Gecko attains 9 cm and is found throughout drier parts of eastern Australia. It is very similar in form and habit to the Eastern Spiny-tailed Gecko *Diplodactylus intermedius* from which it can be distinguished by having four instead of two rows of spines on the tail. The spines exude a sticky fluid when the lizard is alarmed.

A tree-dweller, it is usually located resting under loose bark, in upright hollow logs or occasionally under rocks, during the day.

SPOTTED DTELLA
Gehyra punctata (Fry, 1914)
Fig. 169

Found throughout most of north and north-western Australia, this gecko reaches 6–7 cm. It is most common in rocky habitats, but also enters termite mounds. More than one species is probably being identified here.

Commonly occurring in large population densities, these lizards constitute an important food source for pigmy monitors, small snakes, etc.

Females only lay one egg per clutch, possibly an adaptation to the harsh environment in which these geckoes often live.

Spotted Dtella *Gehyra punctata*

170   Tree Dtella *Gehyra variegata* (Bourke, NSW).

171   *Gehyra* sp. (Lake Argyle, WA).

TREE DTELLA
Gehyra variegata (Dumeril and Bibron, 1836)
Fig. 170

Found throughout most drier parts of Australia, more than one species may be included here. This lizard grows to 7 cm.

It is arboreal along rivers and in forests. In rocky areas this lizard lives among the rock outcrops, sometimes in very high population densities.

This gecko will enter houses and in country towns will commonly be found around street lights feeding on insects at night.

Tree Dtella *Gehyra variegata*

BYNOE'S GECKO
Heteronotia binoei (Gray, 1845)
Figs 172–174

This small 7-cm gecko is highly variable in colour, ranging from whites, reds, browns, greys or even black, with or without flecks, or some kind of colour pattern.

It is found throughout Australia except for the far south-east and far south-west. It is possibly the most

Bynoe's Gecko *Heteronotia binoei*

abundant gecko in Australia, occupying all habitats, and is often found in very large numbers. It is a principal food source for many other species of reptile. Most specimens are found by day hiding under any sort of ground cover. This terrestrial gecko feeds on insects at night.

The female produces two eggs. Parthenogenesis, where all specimens are female and can produce fertile eggs without males, is known in this species. Parthenogenesis appears to result from an unusually high number of chromosomes, in the form of a triploid set, as opposed to the usualy diploid set found in non-parthenogens.

BEADED GECKO
Lucasium damaeum (Lucas and Frost, 1896)
Fig. 175

Found throughout the most arid parts of Australia, this 7-cm gecko is found in all states of mainland Australia.

The Beaded Gecko is a terrestrial species which by day occupies burrows constructed by other animals, concealed in spinifex *Triodia* spp. or other vegetation.

At night this gecko covers considerable areas in search of its insect food, and is commonly found crossing roads.

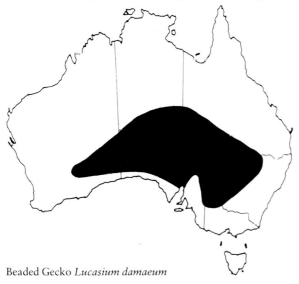

Beaded Gecko *Lucasium damaeum*

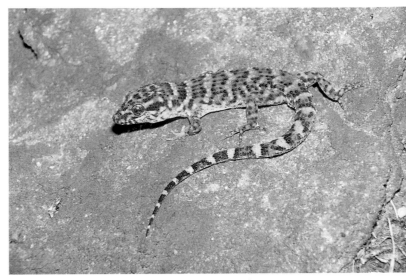

172   Bynoe's Gecko *Heteronotia binoei* (Lake Argyle, WA).

173   Bynoe's Gecko *Heteronotia binoei* (Broome, WA).

174   Bynoe's Gecko *Heteronotia binoei*, head (Boggabri, NSW).

175   Beaded Gecko *Lucasium damaeum* (Bourke, NSW).

176   Rough Knob-tailed Gecko *Nephrurus asper* (Alice Springs, NT).

177   Rough Knob-tailed Gecko *Nephrurus asper*, head (Alice Springs, NT).

178   Centralian Knob-tailed Gecko *Nephrurus laevissimus* (Ayers Rock, NT).

ROUGH KNOB-TAILED GECKO
Nephrurus asper Gunther, 1876
Figs 176, 177

The Rough Knob-tailed Gecko is found throughout drier parts of Queensland, the Northern Territory and north-west Western Australia, including the Kimberleys. This distinctive gecko is usually found in rocky habitats.

One of Australia's largest geckoes, it measures 20 cm and has a robust build. When walking, its body and tail are raised high off the ground.

The Rough Knob-tailed Gecko will stalk and chase its prey, which consists of arthropods, spiders and smaller lizards. It is unable to shed its tail.

Rough Knob-tailed Gecko *Nephrurus asper*

CENTRALIAN KNOB-TAILED GECKO
Nephrurus laevissimus Mertens, 1958
Fig. 178

Distributed in inland parts of South Australia, the Northern Territory and Western Australia, this gecko attains 13 cm. Like all 'Knob-tailed Geckoes', it is recognisable by the distinctive round knob at the end of a short fat tail.

It is restricted to spinifex *Triodia* spp., covered sand

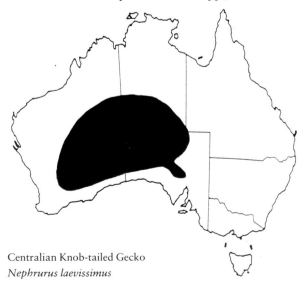

Centralian Knob-tailed Gecko
Nephrurus laevissimus

dunes and plains, where it lives by day in well-concealed burrows. This species feeds at night on small insects and other lizards. Although found in the hottest, driest parts of the country, this gecko has a relatively low tolerance for heat and dehydrates easily. It is most active at temperatures lower than those which other desert species prefer.

SMOOTH KNOB-TAILED GECKO
Nephrurus levis De Vis, 1886
Figs 179, 180

This large, 13-cm gecko is found throughout most parts of arid Australia.

The Smooth Knob-tailed Gecko lives in a variety of habitats but is most common in sandy habitats, where it lives by day in burrows either dug by itself or other animals.

This lizard forages at night, feeding on insects and smaller geckoes. It lays two eggs.

179 Smooth Knob-tailed Gecko *Nephrurus levis levis* (Andamooka, SA).

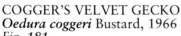
Smooth Knob-tailed Gecko *Nephrurus levis*

180 Pilbara Smooth Knob-tailed Gecko *Nephrurus levis pilbaraensis* (Shay Gap, WA).

181 Cogger's Velvet Gecko *Oedura coggeri* (north of Charters Towers, QLD).

COGGER'S VELVET GECKO
Oedura coggeri Bustard, 1966
Fig. 181

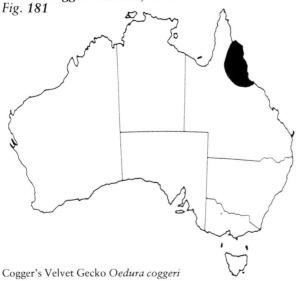

Cogger's Velvet Gecko *Oedura coggeri*

182   Lesueur's Gecko *Oedura lesueurii* (Pearl Beach, NSW).

183   Lesueur's Gecko *Oedura lesueurii*, head (Pearl Beach, NSW).

184   Zig-zag Gecko *Oedura rhombifer* (Port Douglas, QLD).

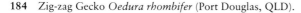

This lizard is found on the eastern side of Cape York Peninsula. It attains 12 cm.

It is most common in rocky habitats where it lives under exfoliations and within rock crevices. This species also has arboreal tendencies. It is sometimes found in small colonies.

At night Cogger's Velvet Gecko feeds on insects present around rock outcrops and adjacent areas. It lays two eggs.

LESUEUR'S GECKO
Oedura lesueurii (Dumeril and Bibron, 1836)
Figs 182, 183

The Lesueur's Gecko reaches 13 cm and is found along the coast and ranges of New South Wales and south-eastern Queensland.

This species occurs in large numbers on sandstone ridges in the Sydney area, where it is usually the dominant gecko species, and the dominant food of the Broad-headed Snake *Hoplocephalus bungaroides*.

In Sydney and elsewhere these lizards are usually caught sheltering under rock exfoliations and crevices where they are preyed on by Tree Snakes and other reptiles. At night they feed on insects adjacent to the rock outcrops. Females lay two eggs.

In some areas these geckoes move into beehive boxes and appear to coexist with the bees without any problems. It is not known if the geckoes are feeding on the bees.

Lesueur's Gecko *Oedura lesueurii*

ZIG-ZAG GECKO
Oedura rhombifer Gray, 1845
Fig. 184

This gecko is found throughout tropical Australia and some arid parts of the Northern Territory and Queensland. It grows to 14 cm.

The Zig-zag Gecko occupies most habitats. It is both ground-dwelling and arboreal, sheltering by day under ground litter and tree bark.

It is common around human habitations, and is often caught searching for insects at night.

Zig-zag Gecko *Oedura rhombifer*

185 Robust Velvet Gecko *Oedura robusta* (Beerwah, QLD).

ROBUST VELVET GECKO
Oedura robusta Boulenger, 1885
Fig. 185

This gecko lives along the coast, ranges and nearby areas of northern New South Wales and Queensland. It attains 18 cm.

This mainly arboreal species is usually found during the day under loose bark on trees or rock crevices in rocky areas. At night it forages close to its daytime abode.

Robust Velvet Geckoes commonly live and breed in human dwellings where they will often lay eggs in cupboards, on shelves behind books, etc. These eggs usually hatch, apparently without problems, under these conditions.

When first caught this gecko will sometimes bite its aggressor, although the bite is only feeble and never even draws blood.

186 Tryon's Gecko *Oedura tryoni* (Moombi Range, NSW).

ranges of north-east New South Wales and south-eastern Queensland. A rock-dwelling species, most specimens are found sheltering under rock exfoliations during the day, often in small groups of up to six individuals. At night specimens can be found foraging on open rock faces in search of their insect diet.

Juveniles are duller in colour than adults.

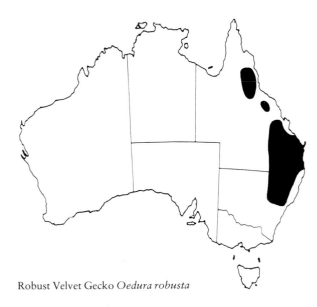

Robust Velvet Gecko *Oedura robusta*

TRYON'S GECKO
Oedura tryoni De Vis, 1884
Fig. 186

The Tryon's Gecko attains 17 cm and is common in

Tryon's Gecko *Oedura tryoni*

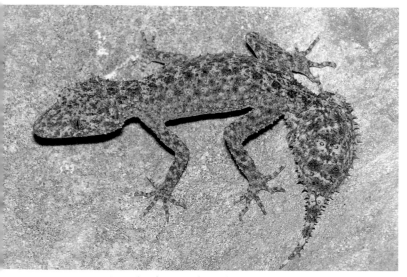

187   Southern Leaf-tailed Gecko *Phyllurus platurus* with original tail (Glenbrook, NSW).

188   Southern Leaf-tailed Gecko *Phyllurus platurus* without tail (West Head, NSW).

189   Southern Leaf-tailed Gecko *Phyllurus platurus* with regenerated tail. Note lack of spines (West Head, NSW).

190   Barking Gecko *Underwoodisaurus milii* with original tail (Boggabri, NSW).

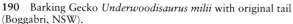

SOUTHERN LEAF-TAILED GECKO
Phyllurus platurus (White, ex Shaw, 1790)
Figs 187–189

This 14-cm gecko is found along the central coast and ranges of New South Wales where it lives in sandstone habitat. Most specimens are found in rock crevices, caves, overhangs and on the underside of loose rocks. Around Sydney this species will enter human habitation where it feeds on flies and other insects. Regenerated tails of this lizard lack the spines found on the original tail (illustrated).

The Southern Leaf-tailed Gecko is commonly confused with the Barking Gecko *Underwoodisaurus milii* because of its habit of making a loud barking sound when caught.

Wild specimens of this gecko usually are found carrying Red Mites (Arachnida), visible as tiny red spots. However, these mites do not seem to harm the geckoes in any way. (They are potentially fatal to many reptiles if present in sufficient numbers.)

Two eggs are usually laid at the back of a rock crevice.

Southern Leaf-tailed Gecko *Phyllurus platurus*

BARKING OR THICK-TAILED GECKO
Underwoodisaurus milii (Bory de Saint-Vincent, 1825)
Figs 190–192

This gecko may exceed 15 cm in length. Found throughout the southern half of Australia, roughly in a line from Brisbane to Perth, this lizard occupies all types of habitat from wet forests to arid scrubs.

This species shelters under most types of ground cover and is often found in small family colonies, of between two and six individuals. Regenerated tails lack the distinctive pattern or shape of the original, being a consistent greyish, purplish-black colour without spines and more rounded in shape.

Two large eggs are produced.

191 Barking Gecko *Underwoodisaurus milii* with regenerated tail (Boggabri, NSW).

Barking Gecko *Underwoodisaurus milii*

192 Barking Gecko *Underwoodisaurus milii* head (Boggabri, NSW).

Family Pygopodidae (Legless Lizards)

Found only in the Australasian area, more than thirty species are currently identified. These lizards have no forelimbs and the hindlimbs are apparent only on close inspection. These are usually only small modified scales and do not aid in locomotion. Legless lizards lack movable eyelids (like snakes) and usually have smooth scales.

Legless lizards can be distinguished from snakes in the following ways:

1 Legless lizards have broad and fleshy tongues, not forked tongues.

2 The ventral scales in legless lizards are usually similar in size to the dorsal scales, while in snakes the ventral scales are always much wider. The ventral scales are always in double rows, not single like in snakes.

3 Unbroken tails in legless lizards are always as long as or longer than the body (as taken from the vent), while the reverse is true for snakes.

4 Most legless lizards have a visible external ear, unlike snakes.

With the exception of the Scalyfoot *Pygopus*

lepidopodus, no Pygopodids exceed a metre in length. All lay two eggs, and can readily shed their tail, which will regenerate rapidly. Diets and habits vary among species.

Males of many (if not all) species possess pelvic spurs in addition to rudimentary hind limbs. These spurs, which are modified spine-like scales, are located beyond the vent, and are covered by the 'hind legs' of pygopodids when these are folded against the body. The spurs are thought to assist the male in gripping the female during copulation.

Delma molleri Lutken, 1863
Fig. 193

Delma molleri attains 30 cm and is one of a number of similar species in the genus *Delma*. *Delma molleri* is restricted to south-eastern South Australia, in the vicinity of Spencer Gulf, where it is fairly common.

In winter most specimens are found sheltering under ground cover such as rocks and logs. During warmer weather specimens are most commonly seen crossing roads at night.

This little-known species is assumed to be insectivorous.

193   *Delma molleri* (Crystal Brook, SA).

Delma molleri

SPINIFEX SNAKE LIZARD
Delma nasuta Kluge, 1974
Fig. 194

One of the more common legless lizards found throughout mainland Australia, the Spinifex Snake Lizard occurs throughout the drier parts of the western two-thirds of the country. It grows to 40 cm long.

This nocturnal species occurs in a variety of grassy habitats where it is assumed to feed on small insects, and possibly small lizards. This species often occurs in one area in large numbers while being absent from similar adjacent habitat.

The breeding habits and general biology are largely unknown.

194   Spinifex Snake Lizard *Delma nasuta* (Halls Creek, WA)

195   *Delma tincta* (Halls Creek, WA).

Spinifex Snake Lizard *Delma nasuta*

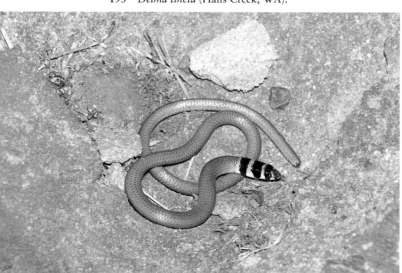

Delma tincta De Vis, 1888
Fig. 195

This 30-cm lizard is found throughout the northern two-thirds of Australia. It usually but not always has dark head markings, which make this lizard superficially similar in appearance to young venomous Brown Snakes

Pseudonaja spp., an adaptation which possibly serves to deter potential predators.

Found in all types of habitats, from rainforests to treeless deserts, this lizard shelters under cover such as rocks, logs, etc., during the day. At night it is commonly seen crossing roads, but is rarely caught then because of its habit of standing up on its tail and jumping away at high speed, like some kind of hopping animal.

Presumed to be insectivorous, this mainly nocturnal lizard is little known.

196 Burton's Legless Lizards *Lialis burtonis*, red and grey forms (Lake Argyle, WA).

Delma tincta

BURTON'S LEGLESS LIZARD
Lialis burtonis Gray, 1835
Figs 196–198

This lizard sometimes exceeds 60 cm and is found in most parts of Australia and New Guinea. The Burton's Legless Lizard is always recognisable by its long, sharp-pointed snout. Along the east coast specimens are either red or grey in base colour. Elsewhere specimens may be either red (including brown), yellow or grey in base colour. The grey specimens may range from very dark to nearly white. Many specimens may have longitudinal markings, particularly white markings around the head and neck region, or along the body.

Burton's Legless Lizard *Lialis burtonis*

197 Burton's Legless Lizards *Lialis burtonis*, two grey specimens (Cottage Point, NSW).

In hotter areas the Burton's Legless Lizard is mainly nocturnal, while in southern and colder areas it is mainly diurnal, but nocturnal on hot summer nights. It occurs in all types of habitat and feeds on smaller lizards. It will ambush lizards while waiting concealed in grass tussocks.

Resting specimens are found hiding in or under most types of ground cover. When handled it may utter a series of loud squeaks.

COMMON SCALYFOOT
Pygopus lepidopodus (Lacepede, 1804)
Figs 199, 200, 531, 532, 562

The Common Scalyfoot is found over most parts of southern Australia, with the exception of very arid regions. It averages 60 cm in length. South-eastern Queensland specimens will, however, regularly exceed a metre in length, making it the largest kind of legless lizard. This lizard also has a prehensile tail.

Colour ranges from greys to reds, with or without markings, and this species is recognisable by its 'rough' keeled scales.

This species occurs in all but the most arid habitats, where it feeds on lizards, insects, which it actively stalks, and some vegetable material. Captive specimens readily

198 Burton's Legless Lizard *Lialis burtonis*, red form (West Head, NSW).

feed on banana. Diurnal in cooler weather, this lizard becomes nocturnal in warm weather.

Eggs are laid in the warmer months and hatch about seventy days later.

199 Common Scalyfoot *Pygopus lepidopodus* (West Head, NSW).

Common Scalyfoot *Pygopus lepidopodus*

200 Common Scalyfoot *Pygopus lepidopodus* (Mount Glorious, QLD).

201 Hooded Scalyfoot *Pygopus nigriceps*, smaller desert form (Goldworthy, WA).

HOODED SCALYFOOT
Pygopus nigriceps (Fischer, 1882)
Figs 201–203

This smaller relative of the Common Scalyfoot *Pygopus ledidopodus* occurs in all parts of Australia except for the wettest parts of the south. It attains 45 to 55 cm and is readily distinguished by its hood covering the nape region.

This species occurs in all types of habitat where it feeds on most types of available insect, small reptile and some plants. Two main colour forms occur. Desert specimens are usually a bright orangish colour and near-desert and other forms are usually slightly larger and often duller in colour, generally being more brownish. (Two species may be represented here.)

Mainly nocturnal due to the generally hot weather where it occurs, southern specimens are diurnal in the cooler months.

Hooded Scalyfoot *Pygopus nigriceps*

202 Hooded Scalyfoot *Pygopus nigriceps*, larger 'tropical' and eastern form (Lake Argyle, WA).

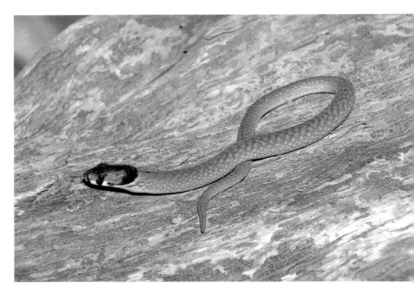

203 Hooded Scalyfoot *Pygopus nigriceps*, juvenile (Bourke, NSW).

204 *Anomalopus mackayi* (Boggabri, NSW).

Family Scincidae (Skinks)

The largest family of lizards in the world, skinks are found in all parts of Australia. About 350 species are currently recognised as occurring within Australia. The 'typical' skink is a smooth-scaled lizard which has well-developed legs and is diurnal. However, many skinks do not fit this pattern. Some have spikey scales, others lack legs or have reduced limbs, while others are nocturnal. Reproductive habits range from egg-laying (oviparous) to live-bearing (ovoviviparous and viviparous), depending on species.

The only positive ways to distinguish a skink are by the head scales, which are usually symmetrical and large, and the presence of a large broad fleshy tongue. Within Australia no other lizards are similar to skinks.

They range in adult size from 5 cm to 60 cm.

Anomalopus mackayi Greer and Cogger, 1985
Figs 204, 205

This is a small skink that has been well known for a number of years but was only recently formally described. It attains 20 cm and is found on the north-western slopes of New South Wales and adjacent parts of Queensland.

205   *Anomalopus mackayi* (Boggabri, NSW).

Most common in rocky and woodland habitats, this species is usually found under ground cover during the day. At dusk it actively forages for the insects it feeds on. This species is an egg-layer.

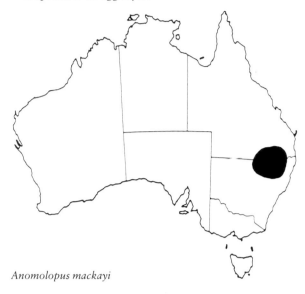

Anomolopus mackayi

Anomalopus swansoni Greer and Cogger, 1985
Fig. 206

This lizard, found along the central coast and adjacent parts of New South Wales, may reach 23 cm in length.

Anomalopus swansoni is usually found in heathland and similar types of habitat. It is a burrowing species that is usually caught when sheltering under bits of ground litter, such as rocks, logs and often man-made rubbish. It lays eggs and feeds on insects.

206   *Anomalopus swansoni* (Port Stevens, NSW).

207   Rainbow Skink *Carlia rhomboidalis* (Port Douglas, QLD).

Anomolopus swansoni

RAINBOW SKINK
Carlia rhomboidalis (Peters, 1869)
Fig. 207

This 15 cm species is common throughout northern tropical coasts of Queensland and offshore islands. It prefers rainforest and similar habitats.

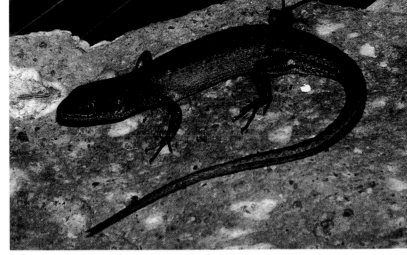

A diurnal species, this swift lizard is usually seen running at high speed over ground cover, where it feeds on insects. It is very hard to catch when active, and will not hesitate to shed its tail when caught.

During the breeding season males have red or occasionally blue throats which tend to fade at other times of the year.

Rainbow Skink *Carlia rhomboidalis*

208 *Carlia vivax* (Hervey Bay, QLD).

Carlia vivax (De Vis, 1884)
Fig. *208*

This 8-cm skink is common along the north coast and adjacent areas of New South Wales, and along the coast and near-coastal parts of Queensland.

It is highly variable in colour, even within a given population or within offspring from a single clutch of eggs.

This small skink occurs in a variety of habitats, where it is diurnal and insectivorous. Little is known of the biology of this species, although in February 1987 I found a pair of these skinks mating under a log at Hervey Bay, Queensland. It is not known when this species lays its eggs.

209 Western Snake-eyed Skink *Cryptoblepharus plagiocephalus* (Boggabri, NSW).

WESTERN SNAKE-EYED SKINK
Cryptoblepharus plagiocephalus (Cocteau, 1836)
Fig. *209*

A small skink, this species only attains 9 cm. Common throughout Australia except for the far east coast and

Carlia vivax

Western Snake-eyed Skink
Cryptoblepharus plagiocephalus

210 Eastern Snake-eyed Skink *Cryptoblepharus virgatus* (St Ives, NSW).

south coastal regions, it is called the Snake-eyed Skink because its lower eyelid is fused and immovable, forming a spectacle covering the eye. Thus its eyes are permanently open, not unlike a snake.

This agile diurnal skink is found in all habitats, where it is usually seen by day around boulders and climbing trees in search of insects. Like all skinks of the genus *Cryptoblepharus* this lizard moves in a distinctive rapid stop/start manner, making it relatively difficult to catch. The Snake-eyed Skink has no hesitation in shedding its tail when threatened.

It is an egg-layer.

EASTERN SNAKE-EYED SKINK
Cryptoblepharus virgatus (Garman, 1901)
Fig. 210

Eastern Snake-eyed Skink
Cryptoblepharus virgatus

211 Leonhard's Skink *Ctenotus leonhardii* (Hervey Bay, QLD).

212 *Ctenotus pantherinus* (Ayers Rock, NT).

Found throughout far southern and eastern Australia, this lizard is similar in most respects to the Western Snake-eyed Skink *Cryptoblepharus plagiocephalus*. The Eastern Snake-eyed Skink *Cryptoblepharus virgatus* is found in most types of habitat.

It shelters under tree bark and in rock crevices from where it searches for insects during the day. In built-up areas it is commonly found on brick walls of houses, etc.

It lays eggs in the warmer months.

LEONHARD'S SKINK
Ctenotus leonhardii (Sternfeld, 1919)
Fig. 211

Common in drier habitats throughout Australia, this 11-cm skink is usually found active during the day foraging, or under ground cover such as fallen logs and tree bark.

An insectivorous species, it appears to be commonly attacked by predators, as many specimens caught in the wild have regenerated tails.

It is an egg-layer whose hatchlings measure about 2–3 cm.

Leonhard's Skink *Ctenotus leonhardii*

Ctenotus pantherinus (Peters, 1866)
Fig. 212

Ctenotus pantherinus attains 25 cm and is found in drier parts of all mainland states except Victoria. In New South Wales it only occurs in the very far north-west. *Ctenotus pantherinus* is larger and more stout than most other *Ctenotus*.

This quick-moving diurnal species is most common in sandy spinifex habitats, but is found in other, similar dry habitats. In some areas this species is found in very high population densities.

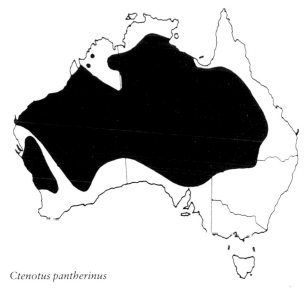

Ctenotus pantherinus

Ctenotus regius Storr, 1971
Fig. 213

Ctenotus regius is found throughout most parts of the eastern two-thirds of South Australia, and adjacent parts of Victoria, New South Wales, Queensland and the Northern Territory.

This 15-cm skink is similar in most respects to Leonhard's Skink *Ctenotus leonhardii*, and is most commonly found in areas with clay and loamy soils and little ground cover. It is very common in grazing country.

213   *Ctenotus regius* (Crystal Brook, SA).

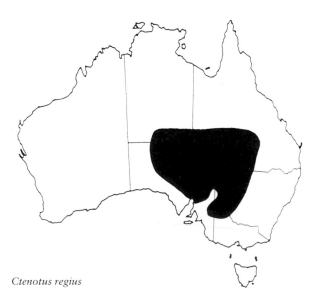

Ctenotus regius

214   Striped Skink *Ctenotus robustus* (Sofala, NSW).

215   Striped Skink *Ctenotus robustus*, head (Boggabri, NSW).

216   Copper-tailed Skink *Ctenotus taeniolatus* (Pearl Beach, NSW).

STRIPED SKINK
***Ctenotus robustus* Storr, 1970**
Figs 214, 215

A 30-cm lizard, the Striped Skink is found widely throughout south-eastern and northern Australia. It occupies all types of habitat, particularly grassy country.

Usually seen during the day in pursuit of its insect food, this swift lizard will often take refuge in a well-concealed, short, shallow burrow when pursued. It is active at very high temperatures and is an egg-layer.

Striped Skink *Ctenotus robustus*

COPPER-TAILED SKINK
***Ctenotus taeniolatus* (White, ex Shaw, 1790)**
Fig. 216

This 18-cm lizard is found in coastal and near-coastal areas of New South Wales and Queensland, and in a small part of far northern Victoria.

This lizard usually occurs in rocky habitats, where it

217   *Ctenotus* sp. (Crystal Brook, SA).

shelters under rock slabs in especially dug burrows. In winter months, large numbers of these lizards can be seen by lifting small rocks on rock outcrops in bush country, particularly around Sydney. This fast-moving lizard may be found during the day in warmer weather in pursuit of its insect food.

Many specimens appear to be an intermediate host for internal parasites that can adversely affect snakes and other animals that eat the lizards, particularly pentastomids (lung parasites).

It lays about five eggs in spring, which hatch some seventy days later. Hatchlings measure about 6 cm.

218 *Ctenotus* sp. (Kunnanurra, WA).

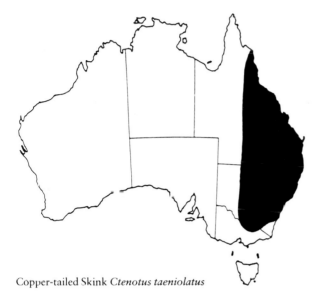

Copper-tailed Skink *Ctenotus taeniolatus*

219 Cunningham's Skink *Egernia cunninghami*, 'typical' form (Cox's River, NSW).

CUNNINGHAM'S SKINK
Egernia cunninghami (Gray, 1832)
Figs 219–223, 482

Possibly more than one species is involved here. This lizard attains 40 cm in length and is found in hilly areas of south-eastern Australia. It is distinguished by the modified scales on its body which form large spines. The colour varies with locality, ranging from plain greys and browns to various combinations of black, red, white,

Cunningham's Skink *Egernia cunninghami*

220 Cunningham's Skink *Egernia cunninghami*, 'typical' form, head (Oberon, NSW).

brown, etc, usually in flecks. Coastal New South Wales and northern specimens appear to be slightly smaller than others.

This species is usually found around large rock outcrops throughout its range where it shelters in rock crevices, or under large slabs of rock. When threatened it

221 Cunningham's Skink *Egernia cunninghami*, 'typical' form, captive bred juvenile of 'mixed' parentage.

222 Cunningham's Skink *Egernia cunninghami* 'New England' form (Tamworth, NSW).

223 Cunningham's Skink *Egernia cunninghami*, 'Sydney sandstone' form (Glenbrook, NSW).

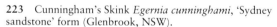

puffs up its body, wedging itself within the rock crevice with its spines, making it almost impossible to dislodge.

The Cunningham's Skink is mainly vegetarian, although it eats large numbers of beetles around December and January.

It is often found in small colonies and family groups ranging from two to ten individuals. Four to six live young are born between February and May. These take about six years to mature.

PYGMY SPINY-TAILED SKINK
Egernia depressa (Gunther, 1875)
Figs 224, 225

Found in most desert and near-desert parts of Western Australia, this lizard attains 17 cm in length.

The Pygmy Spiny-tailed Skink is usually found in association with rock outcrops throughout its range; however, large numbers also inhabit the big termite hills in the Pilbara region.

This lizard also uses its spines to wedge itself in crevices when threatened. The diet is varied, consisting of both insects and plant material. It bears live young.

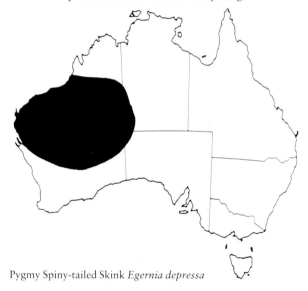

Pygmy Spiny-tailed Skink *Egernia depressa*

224 Pigmy Spiny-tailed Skink *Egernia depressa* (Shay Gap, WA).

MAJOR SKINK
Egernia frerei Gunther, 1897
Fig. 226

The colour of this 45-cm lizard ranges from nearly black through various shades of brown to a light fawn colour. It is found in far north-eastern New South Wales, along the east coast of Queensland, and Arnhem Land in the Northern Territory. It is usually found in thick forest and rainforest habitats. Active during the day, this shy lizard is usually seen in clearings adjacent to its forest habitat.

It feeds on insects, smaller reptiles and various plants, and gives birth to live young.

Major Skink *Egernia frerei*

LAND MULLET
Egernia major (Gray, 1845)
Fig. 227

One of Australia's largest skinks, the Land Mullet attains more than 60 cm. It is so named because of its large shiny fish-like body scales, and the fact that when it moves, it looks a bit like a swimming mullet.

Found in rainforests and similar habitats along the north coast of New South Wales and south-eastern

Land Mullet *Egernia major*

Queensland, this lizard has similar habits to the Major Skink *Egernia frerei*.

It often lives in burrows concealed by Lantana bush in clearings adjacent to rainforests, to which it will retreat when alarmed.

It gives birth to about seven live young in late summer.

YAKKA SKINK
Egernia rugosa De Vis, 1888
Fig. 228

Yakka Skink *Egernia rugosa*

225 Pigmy Spiny-tailed Skink *Egernia depressa*, head (Shay Gap, WA)

226 Major Skink *Egernia frerei* (Fraser Island, QLD).

227 Land Mullet *Egernia major* (Ourimbah, NSW).

The Yakka Skink occurs in scattered localities throughout eastern Queensland. This large, 45-cm skink is little known. It lives in colonies often numbering fifty or more individuals, in sandy or rocky areas. When seen foraging during the day, this shy lizard will rapidly take to cover. Its diet consists of plant material, arthropods and other soft-bodied animals.

Two to five live young are produced, which are much more brightly coloured than the adults.

BLACK ROCK SKINK
Egernia saxatilis Cogger, 1960
Figs 229, 230

Found in mountain country of New South Wales and Victoria, this species attains 30 cm.

This skink inhabits rock crevices and exfoliations in

228 Yakka Skink *Egernia rugosa* (Rockhampton, QLD).

exposed places as well as man-made stone constructions. It is diurnal and is usually seen basking outside the crevices to which it will retreat when disturbed. It is presumed to be mainly insectivorous.

About four live young are produced in summer, measuring 7–8 cm in length.

Black Rock Skink *Egernia saxatilis*

GIDGEE SKINK
Egernia stokesii (Gray, 1845)
Figs 231, 232

This 27-cm skink is found in rock outcrops in desert areas of all mainland states except Victoria.

This species is usually found moving close to boulders, rock crevices, etc., from where it rarely strays. When in rock crevices, this lizard puffs up its body, and wedges itself in the crevices by using its strongly spiked body, making it impossible to dislodge.

It often lives in colonies of up to half a dozen individuals. Its presence can often be determined by its habit of using particular defecating sites, where small piles of faeces accumulate.

About five live young are produced. These measure about 6 cm at birth.

229 Black Rock Skink *Egernia saxatilis* (Zig-zag Railway, NSW).

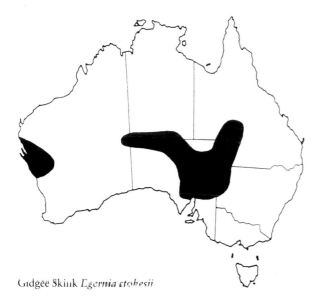

Gidgee Skink *Egernia stokesii*

TREE SKINK
Egernia striolata (Peters, 1870)
Figs 233, 234

This 25-cm lizard is in many respects similar to the Black Rock Skink, from which it can be differentiated by its usually lighter and better-defined coloration. The Tree Skink is found throughout inland eastern Australia and the coast of Queensland.

This species is usually found in trees, under bark, particularly around river flats, and sometimes in large numbers in crevices around rock outcrops, particularly in arid areas. The diet is mixed.

It produces two to three live young, measuring 6 cm.

230 Black Rock Skink *Egernia saxatilis*, head (Zig-zag Railway, NSW).

231 Gidgee Skink *Egernia stokesii* (Broken Hill, NSW).

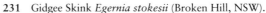

Tree Skink *Egernia striolata*

232   Gidgee Skink *Egernia stokesii*, head (Broken Hill, NSW).

233   Tree Skink *Egernia striolata* (Boggabri, NSW).

234   Tree Skink *Egernia striolata*, head (Boggabri, NSW).

WHITE'S SKINK
Egernia whitii (Lacepede, 1804)
Figs 235–237

Possibly more than one species is involved here. The White's Skink averages 30 cm and is found throughout the coast and ranges of south-eastern Australia, including most of Tasmania. The colour and patterning of this lizard is highly variable, with unmarked and marked specimens occurring within a single population.

Although found in most types of habitats, this lizard prefers rocky areas, where it is usually found under rocks or foraging around rock outcrops. Sometimes two or three specimens of different ages will occupy a single rock with burrows running underneath it, and will retreat to this place when threatened.

This diurnal lizard is mainly insectivorous, and produces three to five live young in summer, measuring 7 cm.

White's Skink *Egernia whitii*

235   White's Skink *Egernia whittii* (West Head, NSW).

BROAD-BANDED SAND SWIMMER
Eremiascincus richardsoni (Gray, 1845)
Figs 238, 561

This lizard is found throughout most drier parts of Australia. It is called a sand swimmer because of its ability to wriggle easily through loose sand.

It is most abundant in, but not restricted to, sandy localities. Generally crepuscular (dusk-active) and nocturnal, this species fossicks for its insect food on the ground surface.

In the Pilbara of Western Australia, I have found large numbers of these lizards inside the giant termite mounds, during the day. It is not known whether these lizards utilise the termite mounds principally for food or shelter, or both. It is also not known whether these specimens leave the mounds every night in search of food elsewhere. Sand Swimmers feed on other lizards which they capture by means of ambush.

Captive specimens become highly excitable when feeding.

This species is little known and is live-bearing.

236 White's Skink *Egernia whittii*, head (Bell, NSW).

Broad-banded Sand Swimmer
Eremiascincus richardsonii

237 White's Skink *Egernia whittii* (Bulls Head, ACT).

238 Broad-banded Sand Swimmer *Eremiascincus richardsonii* (Shay Gap, WA).

239   Alpine Water Skink *Eulamprus kosciuskoi* (Upper Blue Mountains, NSW).

240   Eastern Water Skink *Eulamprus quoyii* (Dural, NSW).

241   Silver Skink *Eulamprus tenuis* (Castle Hill, NSW).

ALPINE WATER SKINK
Eulamprus kosciuskoi (Kinghorn, 1932)
Fig. 239

This 22-cm lizard is found in alpine and mountain areas of New South Wales and adjacent Victoria, at altitudes above 1000 metres.

It is common on the slopes of Mt Kosciusko where it is found active during the day around seepages and small creeks. Elsewhere it occurs in similar habitats, close to water. The Alpine Water Skink is active at temperatures too cold for most other reptiles and is often active around snow drifts or while it is hailing. Although mainly insectivorous, it will feed on some plants.

This lizard is live-bearing and gives birth to about four young in mid to late summer. The gestation period is believed to be about eleven weeks.

Alpine Water Skink *Eulamprus kosciuskoi*

EASTERN WATER SKINK
Eulamprus quoyii (Dumeril and Bibron, 1839)
Fig. 240

This 30-cm lizard is found throughout most parts of New South Wales, coastal Queensland and areas close to the Murray River in Victoria and South Australia.

Eastern Water Skink *Eulamprus quoyii*

The Water Skink is usually found close to creeks and rivers. Only in wetter forests is it likely to stray any distance from water, where it may be found in adjoining farm country, and high and dry rock outcrops. A very swift-moving species, it commonly takes to water when disturbed, often retreating under overhanging creek banks. In suburban areas this species is common in stormwater drains and similar environments.

The Water Skink is omnivorous.

It produces two to five live young in mid summer (January).

SILVER SKINK
Eulamprus tenuis (Gray, 1831)
Fig. 241

Attaining 15 to 20 cm, this lizard occurs along the coast and ranges of New South Wales and Queensland.

Although found in a variety of habitats, the Silver Skink is usually found in wet sclerophyll forest and rainforest habitats. It can be common in built-up areas derived from these habitats, where it inhabits stone walls and lives in crevices in the walls of old houses.

It is particularly common in Sydney's lower North Shore area. In its natural habitat this lizard is often found climbing rock outcrops and trees where it finds its insect food.

It is assumed to be live-bearing but its breeding biology is little known.

Silver Skink *Eulamprus tenuis*

SOUTHERN WATER SKINK
Eulamprus tympanum (Lonnberg and Andersson, 1913)
Figs 242–244

Two species are recognised here. One species, found in the northern tablelands and southern highlands of New South Wales, is called the Warm Temperate Form (WTF). The other species, from most of Victoria and far southeastern South Australia, is called Cold Temperate Form (CTF). In 1984 Wells and Wellington assigned the name *Eulamprus heatwolei* to the Warm Temperate Form. However, their publication was the subject of possible suppression by the International Commission on Zoologi-

242 Southern Water Skink *Eulamprus tympanum* (WTF) (Blackheath, NSW).

243 Southern Water Skink *Eulamprus tympanum* (CTF) (Snowy Mountains, NSW).

cal Nomenclature, so it is uncertain whether the new scientific name will stand.

Both lizards are essentially similar in form and biology, attaining 25 cm in length. They can be differentiated on close inspection by subtle differences in pattern, although both species only occur together around the Snowy Mountains of New South Wales and adjacent areas.

Both species are diurnal and omnivorous, and are often found long distances from water when they occur in wetter habitats. These lizards may become semi-arboreal in habit. They are sometimes found basking on low tree stumps and similar sites.

Several specimens, including some gravid, of the Warm Temperate Form were accidently frozen solid by a colleague of mine. These were thawed out and survived, no worse off for the ordeal.

Mating takes place in spring, with two to five live young being born some three months later, in summer.

244 Southern Water Skink *Eulamprus tympanum* (CTF), juvenile (Warrandyte, VIC).

245 Three-fingered Burrower *Hemiergis decresiensis* (Lithgow, NSW).

246 *Hemiergis graciloides* (Taringa, Brisbane, QLD).

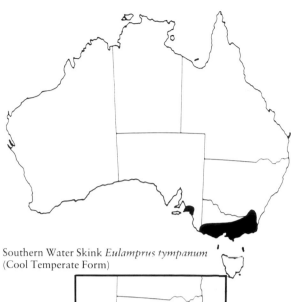

Southern Water Skink *Eulamprus tympanum*
(Cool Temperate Form)

Southern Water Skink *Eulamprus tympanum*
(Warm Temperate Form)

THREE-FINGERED BURROWER
Hemiergis decresiensis (Cuvier, 1829)
Fig. 245

This 13-cm skink is found in highland parts of New South Wales, Victoria and lowland areas in south-eastern South Australia. It is most common in rocky localities, where it is usually found during the day sheltering under rocks, logs and other ground debris. The Three-fingered Burrower has effectively no climbing ability.

Two to five live young are produced in late summer.

Little is known of the species' biology, although it is believed to be nocturnal and insectivorous.

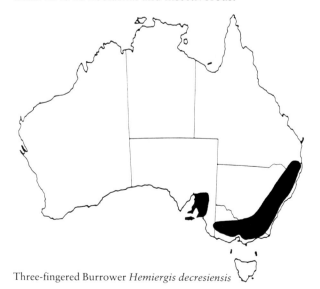

Three-fingered Burrower *Hemiergis decresiensis*

Hemiergis graciloides
(Lonnberg and Andersson, 1913)
Fig. 246

Common around Brisbane and nearby areas, this lizard grows to 8 or 9 cm. This species is typically a forest inhabitant, but readily establishes itself in suburban gardens, where cats commonly catch them and bring them into houses.

This insectivorous lizard only appears to forage above ground at dusk, just before it becomes completely dark. It is assumed to be a burrower the rest of the time.

Hemiergis graciloides

MACCOY'S SKINK
Hemiergis maccoyi (Lucas and Frost, 1894)
Fig. 247

Found in colder parts of far southern New South Wales, the Australian Capital Territory and Victoria, this lizard grows to about 12 cm.

In colder places this species is mainly diurnal, where it is commonly seen in ground litter, apparently feeding on small insects. Elsewhere it is presumed to be mainly crepuscular.

Many specimens are caught under cover in swamps and similar wet places. This species does not seem to be a burrower despite its elongated build.

Maccoy's Skink *Hemiergis maccoyi*

247   Maccoy's Skink *Hemiergis maccoyi* (Bulls Head, ACT).

248   Challenger's Skink *Lampropholis challengeri*, male in front, gravid female at back (North Sydney, NSW).

CHALLENGER'S SKINK
Lampropholis challengeri (Boulenger, 1887)
Fig. 248

This slender lizard grows to 14 cm and is found in moister habitats from Sydney along the coast and near ranges to north Queensland. The southern populations of this lizard might represent a distinct species or sub-species.

It is most common in thick rainforest gullies, where Water Skinks *Eulamprus quoyii* are less common or absent. This is probably due to predation by the Water Skinks.

This diurnal lizard is insectivorous and is usually found during the day under ground litter or foraging. It produces about five eggs in summer.

Challenger's Skink *Lampropholis challengeri*

GARDEN SKINK
Lampropholis delicata (De Vis, 1888)
Fig. 249

This 9-cm lizard is common along the east coast, eastern Victoria, south-eastern South Australia and northern Tasmania.

It is common in suburban back gardens, where it feeds on insects and shelters under any available cover. It is most common in gardens in areas derived from wet sclerophyll forests and rainforests, as opposed to drier habitats, where the Grass Skink *Lampropholis guichenoti* appears to dominate.

It lays two to five eggs in late spring which hatch in mid to late summer. The young lizards mature within twelve months.

Garden Skink *Lampropholis delicata*

249   Garden Skink *Lampropholis delicata* (St Ives, NSW).

250   Grass Skink *Lampropholis guichenoti* (Blacktown, NSW).

GRASS SKINK
Lampropholis guichenoti (Dumeril and Bibron, 1839)
Fig. 250

Common throughout the coast, ranges and slopes of south-eastern Australia, this 9-cm lizard is found in all

Grass Skink *Lampropholis guichenoti*

EL CAMINO COLLEGE
LIBRARY

types of habitat as well as being common in suburban gardens. The Grass Skink is also the dominant small skink in many areas cleared for farming and other human activities.

This insectivorous lizard lays two to five eggs in spring which hatch in summer. Hatchlings measure about 2.5 cm. Sometimes large numbers of females lay eggs in one communal site. More than 200 eggs have been recorded from one such site.

WEASEL SKINK
Lampropholis mustelina (O'Shaughnessy, 1874)
Fig. 251

The 15-cm Weasel Skink lives in the moister parts of the coast and ranges of New South Wales and Victoria.

Although a wet forest inhabitant, this lizard will move into other adjoining habitats, although it tends to remain in shady places. It is common in suburban gardens of Sydney and Melbourne. This lizard is not as fast-moving as most other small skinks.

An insectivorous, diurnal lizard, it lays from two to five eggs in warm moist places, sometimes communally, in spring. They hatch about seventy-five days later. It has been recorded that some specimens deposit their eggs in the same site every year.

251 Weasel Skink *Lampropholis mustelina* (Lane Cove, NSW).

252 Coventry's Skink *Leiolopisma coventryi* (Bulls Head, ACT).

Weasal Skink *Lampropholis mustelina*

COVENTRY'S SKINK
Leiolopisma coventryi Rawlinson, 1975
Fig. 252

This 10-cm lizard lives in the highlands of southern New South Wales and Victoria. It is only readily distinguishable from the Garden Skink *Lampropholis delicata* by the fact that its nostrils are much closer together.

This small lizard is found in association with dead timber and other ground litter in wooded habitats. Largest numbers of this species are found in recently burned forests. This insectivorous lizard is active at very low daytime temperatures. When overwintering this species may be found in a state of torpor in sites, such as within rotten logs, often with a number of other reptiles of the same or other species.

Mating occurs during autumn, winter and spring, with

ovulation occurring about October–November. One to seven live young (average three) are produced in February. Like a number of small skinks, this species has the capability of long-term storage of male sperm.

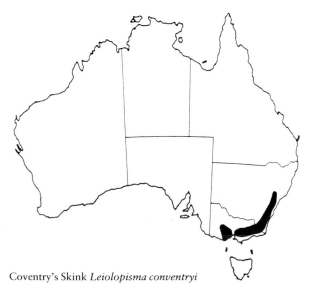

Coventry's Skink *Leiolopisma conventryi*

EL CAMINO COLLEGE LIBRARY

101

253   *Leiolopisma entrecasteauxii* (Form A) (Katoomba, NSW).

254   *Leiolopisma entrecasteauxii* (Form B), male (Bulls Head, ACT).

255   *Leiolopisma entrecasteauxii* (Form B), female (Bulls Head, ACT).

Leiolopisma entrecasteauxii (Dumeril and Bibron, 1839)
Figs 253–255

Two species of lizard are grouped here. They have been termed forms A and B. *Leiolopisma entrecasteauxii* is found throughout the higher country of New South Wales, most of Victoria and Tasmania, and far south-eastern South Australia. These lizards are very common where they occur. The two forms are found in the same area sometimes, but in general Form A is found in warmer habitats than Form B. Both forms are highly variable in colour and patterning. During the breeding season males (usually Form B) may have red shoulders, sides of neck and belly.

Typical specimens of each are illustrated in this book. Form A averages 9 cm while Form B is larger, more heavily built and averages 11 cm. Both lizards are insectivorous, diurnal and live-bearers.

Most specimens are found either under ground litter, or foraging for insects during the day.

Mating takes place in mid to late summer. The sperm is stored internally until ovulation the following spring. After ten to twelve weeks gestation, two to six live young are born around February.

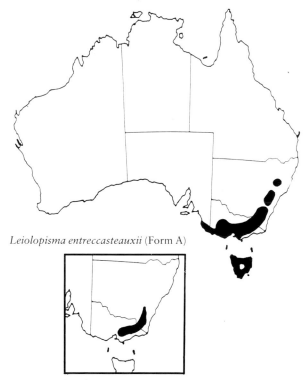

Leiolopisma entreccasteauxii (Form A)

Leiolopisma entreccasteauxii (Form B)

RED-THROATED SKINK
Leiolopisma platynotum (Peters, 1881)
Fig. 256

Found along the east coast and ranges of New South Wales, and adjacent areas, this lizard attains 14 cm. Females are in general by far the larger sex.

Although found in a variety of habitats, this lizard is usually found in association with rock outcrops in rocky areas. It is a very swift diurnal lizard and is usually seen during the day in pursuit of its insect food.

The Red-throated Skink is strictly terrestrial and never climbs vegetation or raised logs.

Red-throated Skink *Leiolopisma platynotum*

256 Red-throated Skink *Leiolopisma platynotum* (Woodford, NSW).

THREE-LINED SKINK
Leiolopisma trilineata (Gray, 1838)
Fig. 257

The Three-lined Skink is common throughout colder parts of south-eastern Australia and south-western Australia. It averages 20 cm. This fast-moving diurnal lizard is usually seen in daytime in search of insects to eat. It prefers grassy woodland habitats where it shelters under logs and other ground litter.

Five to nine soft-shelled eggs are produced in late January.

257 Three-lined Skink *Leiolopisma trilineata* (Bulls Head, ACT).

when found sheltering under ground litter. This lizard feeds on ants, termites and other small insects.

Mating in this species occurs in late summer, autumn and the cooler months. Some populations are oviparous, producing from two to four soft-shelled eggs which hatch shortly after being laid. Other populations, usually in colder areas, are live-bearing and produce two to four young.

Three-lined Skink *Leiolopisma trilineata*

Lerista bougainvillii (Gray, 1839)
Fig. 258

This 15-cm lizard is found in most parts of Victoria, the southern highlands and slopes in New South Wales, and south-eastern South Australia. It is usually found in rocky areas. The taxonomic status of this lizard is in doubt.

A burrowing species, it is usually caught during the day

Lerista bouganvillii

258   *Lerista bougainvillii* (Newnes, NSW).

259   Mueller's Skink *Lerista muelleri* (Boggabri, NSW).

260   Burnett's Skink *Lygisaurus burnettii* (West Head, NSW).

MUELLER'S SKINK
Lerista muelleri (Fischer, 1881)
Fig. 259

This 13-cm lizard is found in most arid parts of Australia and adjoining areas. Found under loose ground litter in the winter months, this species is found in all types of habitat.

During summer it becomes active at dusk, when it may be seen moving around on the ground surface. Mueller's Skink is believed to be insectivorous.

Mueller's Skink *Lerista muelleri*

BURNETT'S SKINK
Lygisaurus burnettii (Oudemans, 1894)
Fig. 260

This small slender lizard grows to 9 cm and is found along the east coast from Sydney, northwards.

It lives in a range of habitats and is most common in moist forested habitats. It is diurnal and feeds on small insects.

Burnett's Skink lays eggs.

Burnett's Skink *Lygisaurus burnettii*

BOULENGER'S SKINK
Morethia boulengeri (Ogilby, 1890)
Figs 261, 262

A 10-cm skink, Boulenger's Skink occurs thoughout most of Queensland, New South Wales, Victoria, South Australia and adjacent areas. Males often have a reddish colour under the neck.

Although found in a variety of habitats, this species is most common in grassy, open woodland habitats, where it feeds on small insects. Its biology is little known.

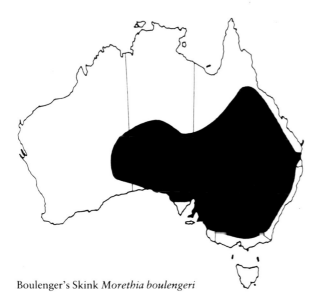

Boulenger's Skink *Morethia boulengeri*

FIRE-TAILED SKINK
Morethia ruficauda (Lucas and Frost, 1895)
Fig. 263

This 8-cm skink is found throughout the north-western quarter of Australia. It occurs in a variety of habitats and is usually seen when active during the day, foraging in ground litter.

This insectivorous species is little known. Similar species are found elsewhere within Australia.

Fire-tailed Skink *Morethia ruficauda*

261 Boulenger's Skink *Morethia boulengeri*, gravid female (Sofala, NSW).

262 Boulenger's Skinks *Morethia boulengeri*, ventral surface. Male is the smaller specimen. (Sofala, NSW).

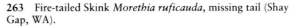

263 Fire-tailed Skink *Morethia ruficauda*, missing tail (Shay Gap, WA).

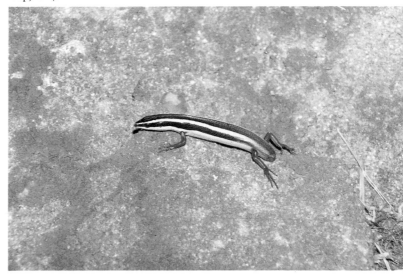

264   Spencer's Skink *Pseudemoia spenceri* (Picadilly Circus, ACT).

265   Three-toed Skink *Saiphos equalis*, gravid female (Galston Gorge, NSW).

266   She-oak Skink *Tiliqua casuarinae* (Katoomba, NSW).

SPENCER'S SKINK
Pseudemoia spenceri (Lucas and Frost, 1894)
Fig. 264

Found in the highland and other very cold areas of Victoria and southern New South Wales, Spencer's Skink attains 14 cm. It occurs in high-altitude forests and woodlands. It is usually found under tree bark, rock exfoliations and crevices, from where it emerges to bask and seek its insect food. A large number of specimens are seen around the walls of bush huts built in this lizard's habitat. It is active at very low temperatures.

Mating takes place in late summer, autumn and spring, with sperm being held by the female until ovulation in late spring. After ten to twelve weeks gestation one to three live young measuring about 3 cm are produced.

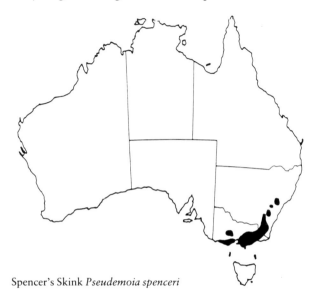

Spencer's Skink *Pseudemoia spenceri*

THREE-TOED SKINK
Saiphos equalis (Gray, 1825)
Fig. 265

A 15-cm lizard, the Three-toed Skink is found along the coast and near ranges of New South Wales and south-eastern Queensland.

Three-toed Skink *Siaphos equalis*

A burrowing lizard, it is usually found under rocks and logs during the day. When disturbed, it often tries immediately to burrow rapidly into any loose soil. This insectivorous lizard is common in suburban back gardens and often is killed in mistake for being a baby snake.

It produces two to four large eggs which hatch within a week or two of being laid.

SHE-OAK SKINK
Tiliqua casuarinae (Dumeril and Bibron, 1839)
Figs 266, 267

Occurring in the coast and near ranges of New South Wales, eastern Victoria and Tasmania, the She-oak Skink attains 30 cm. Larger specimens are usually unmarked, while sub-adults and juveniles usually have distinctive patterned markings on their bodies and dark bands over the head and neck region. Females are by far the larger sex.

Storr and others have proposed that this species should be removed from the Bluetongue genus *Tiliqua* and placed in its own genus *Omolepida*, along with a couple of similar related species.

These lizards are found in a range of habitats, but appear to be most common in grassy grazing country adjacent to forest habitats.

The She-oak Skink when found active makes snake-like movements. Like other skinks with reduced limbs, the limbs are held folded to the side of the body while it engages in snake-like movement. Unlike the Bluetongue lizards, this species will not hesitate to shed its tail if caught. When inactive, this species is usually found hiding under loose ground litter, and is particularly fond of man-made rubbish.

In cold areas, these lizards are diurnal, while in warmer places they tend to be crepuscular.

Mating occurs in October and November, with live young being born some eight to ten weeks later. About five to seven young are produced in December and January. The young measure from 4 to 5 cm.

267 She-oak Skink *Tiliqua casuarinae*, sub-adult (Hazelbrook, NSW).

268 Pink-tongued Skink *Tiliqua gerrardii* (Springwood, NSW).

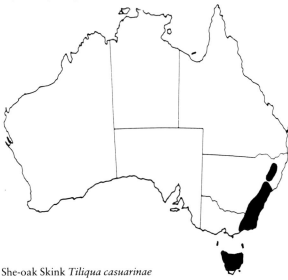

She-oak Skink *Tiliqua casuarinae*

PINK-TONGUED SKINK
Tiliqua gerrardii (Gray, 1845)
Figs 268, 269

The Pink-tongued Skink averages 45 cm and varies in col-

our from patterned to an almost white colour. Juveniles often have distinctive broad bands which usually fade with age. Many Pink-tongued Skinks, particularly young-er specimens, have blue instead of pink tongues. The tail is prehensile.

This species is found from Cape York in Queensland

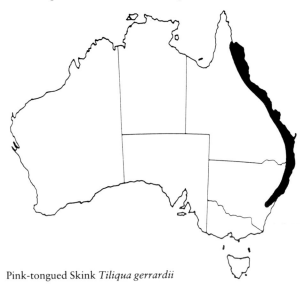

Pink-tongued Skink *Tiliqua gerrardii*

269 Pink-tongued Skink *Tiliqua gerrardii*, new-born (Springwood, NSW).

270 Northern Bluetongue *Tiliqua intermedia* (Arnhem Hwy, NT).

271 Centralian Bluetongue *Tiliqua multifasciata* (Goldsworthy, WA).

and south along the coast and near ranges to Gosford, New South Wales, with an isolated population at Springwood in the Blue Mountains of New South Wales.

This species is most common in wetter forest habitats. It is nocturnal in warm weather and diurnal in colder weather, and feeds almost exclusively on snails and slugs.

About twelve to twenty-five live young measuring 6 cm are produced in summer.

NORTHERN BLUETONGUE
Tiliqua intermedia Mitchell, 1950
Fig. 270

The Northern Bluetongue grows to 60 cm and is found throughout tropical parts of northern Australia. It is closely related to the Eastern Bluetongue *Tiliqua scincoides*. The Northern Bluetongue *Tiliqua intermedia* occurs in various habitats, but is most common in savannah woodland.

It is diurnal and feeds on smaller animals and arthropods, as well as various plant material. It shelters under various types of ground cover, but is particularly fond of sheets of tin commonly found around human habitation.

When disturbed it will stand its ground, puff up its body, open its mouth, and flick its blue tongue.

Females give birth to five to twenty live young during the wet season.

Northern Bluetongue *Tiliqua intermedia*

CENTRALIAN BLUETONGUE
Tiliqua multifasciata Sternfeld, 1919
Figs 271, 272

This lizard averages 40 to 45 cm and occurs in desert areas of Australia and tropical parts of north-western Australia.

Most specimens can be seen crossing roads when they are foraging for food. In the hotter months this lizard is mainly crepuscular and nocturnal, while it becomes day-active in the colder months. It is believed to feed on wild-flowers, small vertebrates and insects.

About two to five live young measuring 10 cm are produced.

Centralian Bluetongue *Tiliqua multifasciata*

BLOTCHED BLUETONGUE
Tiliqua nigrolutea (Quoy and Gaimard, 1824)
Figs 11, 273–277

This lizard attains 60 cm and is found in the highland regions of central and southern New South Wales, most of Victoria, Tasmania and the far south-east of South Australia. The colour varies greatly with location. New

272 Centralian Bluetongue *Tiliqua multifasciata*, juvenile (Victoria River, NT).

273 Blotched Bluetongue *Tiliqua nigrolutea*, NSW alpine form (Zig-zag Railway, NSW).

274 Blotched Bluetongue *Tiliqua nigrolutea*, NSW alpine form, head (Zig-zag Railway, NSW).

275 Blotched Bluetongue *Tiliqua nigrolutea*, NSW alpine form, juvenile (Katoomba, NSW).

South Wales specimens are usually black with bright orange or yellowish blotches, while Victorian and Tasmanian specimens are a much duller bluey-grey. Northern specimens grow larger than their southern counterparts.

These robustly built lizards are omnivorous, although around December in some areas they feed almost exclusively on Christmas beetles.

From four to ten very large young, measuring 15 cm, are born in summer. Occasionally in the wild this species hybridises with the Common Bluetongue *Tiliqua scincoides*. The young produced are intermediate in colour between the two species, and are themselves apparently fertile.

276   Blotched Bluetongue *Tiliqua nigrolutea*, southern form (Monbulk, VIC).

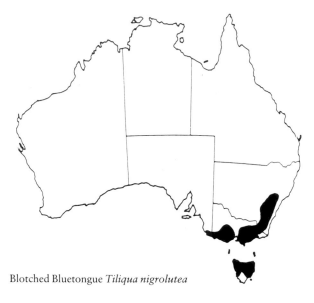

Blotched Bluetongue *Tiliqua nigrolutea*

277   Blotched Bluetongue *Tiliqua nigrolutea*, southern form, head, sub-adult (Monbulk, VIC).

WESTERN BLUETONGUE
Tiliqua occipitalis (Peters, 1863)
Fig. 278

The Western Bluetongue measures about 50 cm. It is found throughout drier parts of southern Australia, from the West Australian coast to inland New South Wales and Victoria, as well as in southern parts of the Northern Territory.

Although found in a variety of habitats this lizard is most common in mallee and spinifex *Triodia* habitats. In some southern areas it is sympatric with the usually more common Eastern Bluetongue *Tiliqua scincoides*, while in some centralian spinifex habitats this lizard occurs together with the Centralian Bluetongue *Tiliqua multifasciata*.

The diet consists mainly of berries, arthropods and spiders.

Records show that five to ten live young are produced.

278   Western Bluetongue *Tiliqua occipitalis* (Renmark, SA).

Western Bluetongue *Tiliqua occipitalis*

279   Eastern Bluetongue *Tiliqua scincoides*, coastal NSW form (St Clair, NSW).

280   Eastern Bluetongue *Tiliqua scincoides*, inland form, which lacks a distinctive temporal stripe. Juvenile. (Bourke, NSW).

EASTERN BLUETONGUE
Tiliqua scincoides (White, ex Shaw, 1790)
Figs 279–281

Found in a number of habitats in south-eastern Australia, the Eastern Bluetongue is probably Australia's best known skink. It grows up to 60 cm, though averaging 45 cm. It is usually but not always greyish in colour, and has a distinctive appearance and pattern. More than one species may be included here. Specimens from west of the Great Divide are sufficiently different from those in coastal and adjacent areas to be at least a different subspecies. (Some obvious external differences are illustrated.)

When disturbed this lizard will often stand its ground, hissing loudly, puffing up its body and holding its mouth agape to reveal its blue tongue. Although this lizard appears to lack well-defined teeth, it is capable of delivering a powerful bite. When biting humans, this species has the habit of not letting go, adding to the pain inflicted.

Where ground cover is thick, this short-limbed species will often move in a snake-like manner, resulting in many specimens being killed in mistake for snakes, particularly Death Adders *Acanthophis antarcticus*.

The Eastern Bluetongue feeds on most types of animal material small enough to be eaten, as well as plant ma-

terial. This diurnal lizard commonly lives in suburban back gardens and is commonly kept as a pet. It is hardy in captivity, becoming completely docile and thriving on bananas and canned petfood.

It gives birth to an average of twelve, 10-cm live young in summer.

281   Eastern Bluetongue *Tiliqua scincoides*, north Queensland form (Cairns, QLD).

Eastern Bluetongue *Tiliqua scincoides*
(All forms currently grouped in this species)

SHINGLEBACK LIZARD
Trachydosaurus rugosus Gray, 1825
Fig. 282

Also called 'Bog Eye', Pine Cone Lizard, Sleepy Lizard, Bobtail and Cow Turd Skink, this very heavily-

282   Shingleback Lizard *Trachydosaurus rugosus* (Dalby, QLD).

built lizard attains 45 cm. Its colour is highly variable, but this lizard is always recognisable by its pine-cone-like scales (shingles). Specimens from west of the Nullabor Plain usually have reddish heads.

This species occurs thoughout drier parts of Queensland, New South Wales and Victoria, and is also found in southern parts of South Australia and Western Australia. Eastern specimens are usually black, white, yellow, brown or combinations thereof. Western specimens might be any of the preceding colours as well as often having reds and oranges, particularly around the head.

The large tail is used as a food and moisture store during periods of food shortage. It cannot be shed.

This slow-moving lizard is commonly found crossing roads during the day.

It feeds on insects, snails, carrion and vegetable material including flowers. It has a preference for yellow flowers. When inactive this lizard shelters under all types of ground cover, and is particularly fond of sheet metal. Consequently many specimens are found around rubbish tips.

When harassed the Shingleback will adopt its typical defensive behaviour. It will inflate its body and expel air in loud hisses, simultaneously holding its mouth agape revealing the broad greyish-blue tongue.

The Shingleback cannot survive in areas of high humidity for long periods, and consequently most specimens placed in captivity in Sydney, Melbourne and Brisbane tend to die of various ailments within short periods.

This lizard produces one or more, usually two, very large live young which develop through a very well-developed placental system, typical of other *Tiliqua*. They measure some 16 cm at birth.

Shingleback Lizard *Trachydosaurus rugosus*

Family Varanidae (Goannas/ Monitor Lizards)

About thirty-six species occur throughout warmer parts of the 'old world' and about twenty-five of these occur in Australia. They are called monitors, or goannas, the terms being interchangeable.

These lizards are all of the typical lizard form and are usually medium-sized to very large lizards. The Komodo Dragon, the world's largest lizard, is a monitor, and

283 Sand-dwelling Skink (species unknown) (Shay Gap, WA).

284 Ridge-tailed Monitor *Varanus acanthurus* (Shay Gap, WA).

Australia's largest lizards belong to this family. These Australian types may exceed 2 metres in length.

All monitors have well-developed teeth and larger ones can give a nasty bite. Goannas are the only Australian lizards to possess a forked tongue.

All types are strictly carnivorous, feeding on insects and other animals, depending on the size of the monitor. When well fed, the base of the tail often becomes thickened as a fat store, to be used in lean times.

All Australian monitors are placed in the genus *Varanus*, although they are sometimes placed within two sub-genera, one group with vertically compressed tails (*Varanus*) and another with tails that are round in section (*Odatria*).

Monitors are egg-layers. The eggs of most species take from three to twelve months to hatch.

RIDGE-TAILED MONITOR
Varanus acanthurus Boulenger, 1898
Fig. 284

This lizard is common thoughout drier parts of the northern half of Australia. It attains 60 cm.

The Ridge-tailed Monitor is found in a variety of habitats, but is most common in rocky locations. It hides in rock crevices where its tail, which has circular ridges along it, will make it hard for specimens to be dislodged. In the Pilbara region of Western Australia, I have caught

113

285   Perenty *Varanus giganteus* (Winton, QLD).

286   Perenty *Varanus giganteus*, head (Winton, QLD).

287   Pygmy Mulga Monitor *Varanus gilleni* (locality unknown).

adults inside termite mounds during the day, in both rocky and flat areas.

Apparently this monitor prefers to feed on smaller lizards, mainly geckoes and skinks, but young specimens are presumed to be mainly insectivorous.

Ridge-tailed Monitor *Varanus acanthurus*

PERENTY
Varanus giganteus (Gray, 1845)
Figs 285, 286

The longest monitor found in Australia, this species grows to more than 2 metres in length. Specimens of nearly 3 metres are known. Found in arid parts of Queensland, South Australia, the Northern Territory and Western Australia, this lizard is usually found in the vicinity of rocky hills and outcrops. It often digs burrows under boulders, fallen trees and other cover, and when pursued will take to these burrows or nearby rock crevices, from where it can be hard to dislodge. Pursued specimens have also been known to take to trees.

Specimens occasionally stray further away from their usual habitat and are sometimes seen crossing roads, mainly in spring and autumn, or in summer during the early morning before it gets excessively hot. The Perenty feeds on a variety of vertebrates and carrion.

Perenty *Varanus giganteus*

In many areas the Perenty actually hibernates during the colder winter months, remaining within its burrow.

About nine eggs are laid, which take from six to nine months to hatch, during the warmer months.

PYGMY MULGA MONITOR
Varanus gilleni Lucas and Frost, 1895
Fig. 287

The Pygmy Mulga Monitor attains 30 cm and is found in drier parts of South Australia, southern Northern Territory, north-western Western Australia and Queensland near Birdsville.

An arboreal species, it is usually found in association with mulga and desert oaks or, less frequently, gum trees. Specimens retreat to cracks in trees or under bark when disturbed foraging. Diet consists principally of small lizards and some insects.

During warm weather this lizard is mainly crepuscular, being active during early morning and dusk.

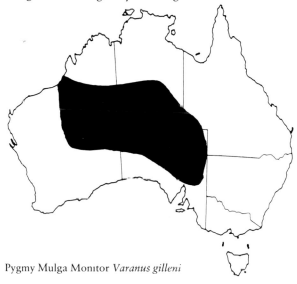

Pygmy Mulga Monitor *Varanus gilleni*

SAND GOANNA
Varanus gouldii (Gray, 1838)
Figs 288–293

Several species are involved here. The Sand Goanna is also known as Gould's Monitor in the eastern states, Bungarra in north and western Australia, and the Racehorse Goanna in the South-west of Western Australia. The name Racehorse Goanna comes from the high speed at which it can run, and it is found in all but the very coldest parts of mainland Australia. These lizards range in adult size from 1 to 2 metres, and pattern and build vary from locality to locality. The largest specimens occur in the vicinity of the Gulf of Carpentaria, Queensland and the Northern Territory. Juveniles are usually much brighter than the adult specimens in colour, particularly in coastal New South Wales, south-eastern Queensland and south-western Western Australia, where juveniles have bright reddish markings. A few distinct 'forms' of this goanna are illustrated here.

This lizard is found in all types of habitat, but is most common in sandy types of area. This ground-dweller forages widely in search of its animal food, which includes smaller reptiles, frogs, birds, etc. It will often dig burrows under large boulders, logs and other ground cover under which it will shelter when not active.

288 Sand Goanna *Varanus gouldii gouldii*, male, coastal NSW form (West Head, NSW).

289 Sand Goanna *Varanus gouldii gouldii*, male, head, coastal NSW form (West Head, NSW).

290 Sand Goanna *Varanus gouldii flavirufus* northern Australian form (Barkly Tableland, NT).

When cornered adult specimens (particularly the northern inland form) will raise the front part of their body off the ground, in a kangaroo-like position and hiss loudly.

Some Aboriginals who catch this lizard for food will throw the dead lizards on to a fire and eat them when the skin begins to peel. The best meat is on the base of the tail and tastes similar to chicken.

Male specimens commonly extrude their hemipenes when caught.

291   Sand Goanna *Varanus gouldii flavirufus* northern Australian form, head (Barkly Tableland, NT).

292   Sand Goanna *Varanus gouldii* sub sp., inland southern form, juvenile (Parkes, NSW).

293   Sand Goanna *Varanus gouldii* sub sp., inland southern form, juvenile, head (Parkes, NSW).

Sand Goanna
Varanus gouldii (All Forms)

MANGROVE MONITOR
Varanus indicus (Daudin, 1802)
Fig. 294

The Mangrove Monitor attains 1.5 metres and is found in mangrove and riverine habitats of Cape York, Queensland and coastal Northern Territory. This very distinctive lizard also occurs in New Guinea and other islands to the north of Australia.

Most specimens can be found when actively in search of their vertebrate food. When threatened the Mangrove Monitor usually takes to water where it will remain until danger passes.

Arnhem Land Aboriginals capture this lizard for food by canoeing down estuarine rivers and beating overhanging mangroves with long poles as they progress. The startled Monitor is then caught as it makes for the water.

Mangrove Monitor *Varanus indicus*

294 Mangrove Monitor *Varanus indicus* (Iron Range, QLD).

Varanus kingorum Storr, 1980
Fig. 295

This 50-cm lizard is common throughout the Kimberley Ranges, Western Australia and nearby parts of the Northern Territory. One of its distinguishing features is its slim build and very long tail, which may be twice as long as its body.

This lizard is usually found around rock outcrops where it feeds on various smaller animals. Specimens can be found under exfoliations where they hide when disturbed. Although only a small monitor, it has sharp teeth and will not hesitate to give a relatively sharp painful bite when caught.

Varanus kingorum

295 *Varanus kingorum* (Kunnanurra, WA).

296 Mertens' Water Monitor *Varanus mertensi* (Hughenden, QLD).

MERTENS' WATER MONITOR
Varanus mertensi Glauert, 1951
Figs 12, 296

This lizard attains 1.2 metres in length. It is found throughout tropical and adjacent parts of Northern Australia, except for eastern Cape York. Populations in some areas have been devastated by the newly introduced Cane Toad *Bufo marinus*.

The Mertens' Water Monitor lives around rivers, lagoons and large creeks, into which it will retreat when threatened. It will often dive into the water and swim away, hiding under submerged rocks, overhanging banks and other well-concealed places. The Mertens' Water Monitor can also remain submerged, with only its nostrils protruding above the water surface, if necessary. This lizard is carnivorous and feeds on available frogs, smaller reptiles, fish, etc. This lizard is known to walk along river bottoms, in a manner not unlike other monitors on land.

About nine eggs are laid in a carefully constructed nesting chamber at the end of a burrow. Leaf litter may be deposited with the eggs by the female, who then seals the burrow and nesting chamber. The eggs hatch between six and nine months later.

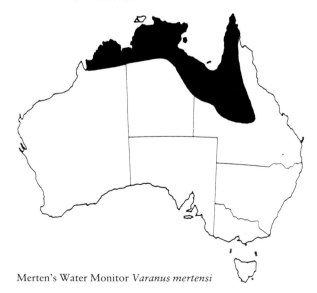

Merten's Water Monitor *Varanus mertensi*

STORR'S MONITOR
Varanus storri Mertens, 1966
Fig. 297

The Storr's Monitor attains 30 cm and is very common around Townsville and Charters Towers, Queensland. It also occurs in scattered localities in north-western Queensland, the Northern Territory and the Kimberleys in Western Australia.

Although primarily an inhabitant of rock outcrops in dry forest and savannah, this lizard will move into adjacent habitats when suitable cover is provided by human development. The Storr's Monitor is common in Charters Towers cemetery where it lives around tombstones and rock gardens.

Feeding on insects and other small reptiles, in late spring this lizard produces about four to six eggs which hatch about four months later. It is an indictment of Australian herpetology that the first specimens of this species to be bred in captivity were bred by an American, Richard Bartlett of Florida.

In the wild, females appear to outnumber males at a ratio of about two to one.

SPOTTED TREE MONITOR
Varanus timorensis Gray, 1831
Figs 298, 299

This 70-cm lizard is found throughout tropical parts of Northern Australia. Its colour is highly variable, with younger specimens often being much more brilliantly coloured than the adults.

The Spotted Tree Monitor lives in various habitats but is most common in savannah woodland habitat. This tree-dweller is usually seen moving around tree trunks and branches where it feeds on insects, geckoes, etc. If found foraging on the ground it will run up the nearest tree. When resting, the Spotted Tree Monitor is found in hollow logs and hollow tree branches, or under loose bark.

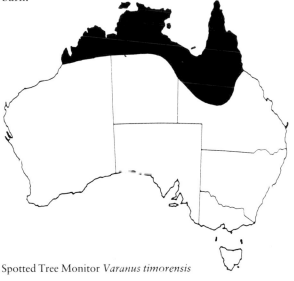

Spotted Tree Monitor *Varanus timorensis*

298 Spotted Tree Monitor *Varanus timorensis* (north QLD).

299 Spotted Tree Monitor *Varanus timorensis*, head (north QLD).

Storr's Monitor *Varanus storri*

297 Storr's Monitor *Varanus storri* (Charters Towers, QLD).

300   Black-headed Monitor *Varanus tristis tristis* (Hughenden, QLD).

301   Black-headed Monitor *Varanus tristis tristis*, head (Hughenden, QLD).

302   Freckled Monitor *Varanus tristis orientalis* (Kingoonya, SA).

BLACK-HEADED MONITOR AND FRECKLED MONITOR
Varanus tristis (Schlegel, 1839)
Figs 300–303

Both monitors are different forms of the same species. The Black-headed Monitor *Varanus tristis tristis* is found in drier parts of western and central Australia, while the Freckled Monitor *Varanus tristis orientalis* is found in north-eastern parts of Australia. Juveniles of *Varanus tristis tristis* often have a brownish head, similar to that of *Varanus tristis orientalis*. Both monitors attain between 60 and 80 cm in length, although the Black-headed Monitor *Varanus tristis tristis* is usually larger than its north-eastern relative.

Both lizards tend to be secretive. They are mainly arboreal, except in very arid areas where they tend to be found around rocky outcrops. Specimens are not often seen unless looked for, and even then they usually make for very inaccessible cover, when seen or pursued.

The diet of these lizards consists mainly of other reptiles, although other animals are sometimes eaten.

Black-headed and Freckled Monitor
Varanus tristis

LACE MONITOR
Varanus varius (White, ex Shaw, 1790)
Figs 304–308, 563

The Lace Monitor is one of Australia's largest lizards with some large males sometimes exceeding 2 metres in length. Specimens are usually a greyish black in colour with thin irregular bands crossing the body. There is, however, a colour form which is common in most drier areas where the colour consists of distinct broad yellow and black bands. These lizards also have a different head morphology from the other specimens, with which they regularly interbreed. I have had it suggested to me that the cause of the banded colour phase is a sex-linked gene, that only expresses itself in males, but further investigation is required here. In general, intermediate colour phases do not occur, adding credibility to the theory given, although for a number of years I have held a female of apparently intermediate colour.

The Lace Monitor occurs in wooded areas of eastern Australia from north Queensland through New South Wales, Victoria and far south-eastern South Australia.

This lizard is both a ground- and tree-dweller, but it will immediately run up the closest vertical object when disturbed. The Lace Monitor has been known to run up a human, when the person had been standing still.

Despite its size the Lace Monitor is a very shy lizard in the wild. It feeds on all forms of carrion and can go for very long periods without food if necessary. In cattle country where food is apparently fairly abundant some specimens become extremely obese.

The males are by far the larger sex, and seem to be the most common in the wild. During the spring mating season, males will fight by clawing and biting, sometimes even mortally wounding one another. Most large males carry many scars from previous fights. These fights often appear to be initiated by the presence of a female close to the males, who otherwise do not seem to be territorial.

About six to twelve eggs are laid in termite mounds, usually in trees.

303 Freckled Monitor *Varanus tristis orientalis*, head (Kingoonya, SA).

Lace Monitor *Varanus varius*

304 Lace Monitor *Varanus varius*, female (St Ives, NSW).

305 Lace Monitor *Varanus varius*, male, juvenile, broad-banded form (Bundaberg, QLD).

306 Lace Monitor *Varanus varius*, old adult male, broad-banded form (Lightning Ridge, NSW).

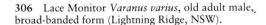

307 Lace Monitor *Varanus varius*, old adult male, broad-banded form, head (Lightning Ridge, NSW).

308 Lace Monitor *Varanus varius*, female. Colour phase is intermediate between the broad-banded and 'typical' patterns. (Cannowindra, NSW).

Order Squamata (Snakes and Lizards)

Suborder Serpentes (Snakes)

Family Typhlopidae (Blind or Worm Snakes)

These snakes are found throughout the world and in most parts of Australia. They are often called Worm Snakes because of their worm-like appearance. All are pinkish or brown in colour, and covered in smooth shiny scales. They are also called Blind Snakes because their eyes are reduced to small spots which give them nominal eyesight. They lack a distinct head and the tail (which is the same thickness as the body) terminates in a blunt spine that aids in progression.

Because of their secretive nature little is known about most types of Blind Snakes and they are without doubt the least known of Australia's snakes. About thirty species are currently recognised in Australia but the means of defining some of them have been dubious, to say the least.

Blind Snakes have traditionally been identified by variations in the cleft present on the nasal scale (which is now known to vary significantly within a single species), body scale counts and general morphology. However, the identification of given species remains difficult because so little is known about most species.

Blind Snakes are for a number of reasons regarded as being separate from all other snakes, and some scientists have placed them somewhere between lizards and snakes. A lizard-like characteristic of Blind Snakes is that their belly scales (ventrals) are a similar size to their dorsal scales, while other snakes have their characteristic broad belly (ventral) scales. Furthermore, Blind Snakes appear to be unable to swallow large prey items by dislocating their jaws in the normal snake-like manner.

Blind Snakes are mainly burrowers, only moving about on the ground surface on warm and/or wet nights. When caught most Blind Snakes release a pungent smell from well-developed anal glands. This smell serves the dual purposes of defence and attracting the opposite sex during the mating season.

Blind Snakes have ventrally positioned shark-like mouths, and are believed to feed on ants and termites. They are the only Australian snakes known to be insectivorous. They are unable to bite humans and lack venom glands.

They are all thought to be egg-layers.

COMMON EASTERN BLIND SNAKE
Ramphotyphlops nigrescens (Gray, 1845)
Fig. 309

Found throughout the coast, ranges and near slopes of Victoria, New South Wales and south-eastern Queensland, this is one of the more common Blind Snakes in this part of the country. It sometimes exceeds 60 cm and is also one of the largest Blind Snakes. It is usually pinkish in colour.

Most specimens are seen during the day under ground cover, or on wet nights crossing bush roads.

Blind Snake *Ramphotyphlops nigrescens*

Ramphotyphlops proximus (Waite, 1893)
Fig. 310

This large species may also exceed 60 cm in length. It is found throughout most of New South Wales and adjoining areas, and the coast and ranges of Queensland up to Cape York. It is usually brownish in colour.

In some texts it has been reported as being very common in built-up parts of Sydney, where it supposedly feeds on insects that accumulate on the bottom of street light poles at night. I have not seen this in many years of living in Sydney.

Like all Blind Snakes, this species is in urgent need of further research.

309 Blind Snake *Ramphotyphlops nigrescens* (Cottage Point, NSW).

Blind Snake *Ramphotyphlops proximus*

310 Blind Snake *Ramphotyphlops proximus* (Terry Hills, NSW).

Family Boidae (Pythons and Boas)

Sub-family Pythoninae (Pythons)

Unlike the Boas, which are mainly confined to the Americas, the Pythons are restricted to Africa, Asia and Australasia. Pythons and Boas are the largest snakes in the world, and Pythons are the largest Australian snakes.

Python taxonomy is currently in a state of dispute. The scientific names used in the following text correlate with the consensus view of Australian herpetologists at the time of writing, and it is highly likely that different names will be in use in the future. (See section on 'Problems of Python classification and Hybrid Pythons' on p. 137.)

Pythons are in general well known. They are slow-moving, heavy-bodied snakes that kill their food by constriction, and although they lack venom glands, larger specimens can give a nasty bite. There are more than a dozen recognised species in Australia, ranging in size from the 60-cm Ant-hill Python *Bothrochilus perthensis* to the 4-m-plus Scrub Python *Morelia amethistina*.

Pythons possess pelvic spurs, which appear to be remnants of hind legs, around the vent. These have the appearance of being nothing more than a modified scale. In some types of Python these spurs are larger in males than females. They are used when mating. The male Python erects the spurs and digs them into the female while moving over her when attempting to copulate.

As opposed to Boas, which are live-bearers, all Pythons lay eggs, which they coil around and incubate until they hatch. Some Pythons can apparently spasmodically twitch their muscles, or 'shiver', and increase the temperature of their bodies and eggs, when incubating the eggs, as is

necessary. Some Pythons have been known to raise the temperature of their eggs by as much as 6°C above that of the ambient air temperature.

Most Python eggs take two or three months to hatch.

With the exception of the Black-headed Python *Aspidites melanocephalus* and the Woma *Aspidites ramsayi*, all Australian pythons possess heat sensitive pits in the labial scales, below the lower jaw. Tests have shown these pits to be sensitive to temperature variations of as little as one-tenth of 1°C. The pits presumably aid these Pythons in their location of warm-blooded prey.

BLACK-HEADED PYTHON
Aspidites melanocephalus (Krefft, 1864)
Figs 311, 312

This distinctive, heavy-bodied snake attains 2 m or more in length and is found throughout tropical and adjacent dry parts of Australia.

The scalation is smooth with 50–65 mid-body rows, 315–355 ventrals, single anal and 60–75 mainly divided subcaudals.

Found in all types of habitat, this ground-dwelling snake is usually encountered moving about at night. Although it will hiss a lot and raise its head when ambushed, it is inoffensive and will rarely bite. It feeds on mammals, birds, lizards and snakes, including venomous types.

In the Pilbara region this species is commonly found inside termite mounds, where it feeds on marsupial mice *Antechinus* spp., and seeks refuge from the daytime heat.

Eggs are thought to be laid about December. About ten are produced. The hatchlings measure about 50 cm long.

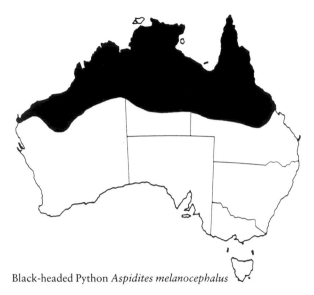

311   Black-headed Python *Aspidites melanocephalus* (Shay Gap, WA).

Black-headed Python *Aspidites melanocephalus*

WOMA
Aspidites ramsayi (Macleay, 1882)
Figs **313, 314,** 536

The Woma is found in many parts of Australia, in all forms of desert habitat, ranging from saltbush plains and mulga scrubs to spinifex plains. It averages 2 metres in

312   Black-headed Python *Aspidites melanocephalus*, partially albinistic specimen, male (The Tits, WA).

length. Young specimens often have black markings on their heads, while Queensland specimens have dark markings around the eyes and are slightly greyer in general colour.

This snake is similar in habit to, and closely related to, the Black-headed Python, although they do not appear to interbreed in the wild. In parts of the Pilbara, Western Australia, both species occur together. Populations of Womas in southern Western Australia have declined sharply in recent years. Some people call the Woma the Sand Python, because of its preference for sandy habitats, and its habit of occupying other animals' burrows in these areas.

The scalation is smooth with 50–65 mid-body rows, 280–315 ventrals, single anal and 40–55 mainly single subcaudals.

These snakes feed on reptiles, mammals, etc., and lay about nine eggs in December. Hatchlings measure 50 cm.

WHITE-LIPPED OR D'ALBERT'S PYTHON
Bothrochilus albertisii Peters and Doria, 1878
Figs 315, 316

The 2–3-metre White-lipped Python is found on northern Torres Strait islands which constitute Australian territory. This species is also found in southern New Guinea, and a smaller similar form is found in the north of New Guinea. The snake pictured is typical of 'Australian' specimens although more brownish-coloured specimens occur, particularly as one moves northwards in New Guinea.

The scalation is smooth with 45–55 mid-body rows, 260–90 ventrals, single anal and 60–80 divided subcaudals.

This terrestrial species occurs in forested habitats and feeds principally on small mammals.

313　Woma *Aspidites ramsayi*, female (Tea Tree, NT).

Woma *Aspidites ramsayi*

314　Woma *Aspidites ramsayi*, juvenile (Charleville, QLD).

315 White-lipped Python *Bothrochilus albertisii* (locality unknown).

316 White-lipped Python *Bothrochilus albertisii*, head, (locality unknown).

Although not often kept in captivity, keepers who have held White-lipped Python rarely had difficulty in breeding this species. It lays about ten eggs which take about sixty days to hatch.

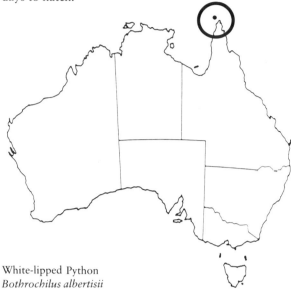

White-lipped Python
Bothrochilus albertisii

CHILDREN'S PYTHON
Bothrochilus childreni Gray, 1842
Figs 317–319

Found throughout tropical Australia, west of Cape York, this snake is highly variable in colour. Specimens may be patterned or unpatterned. Body form and temperament of specimens varies widely between and within localities. Specimens rarely exceed 1.5 metres.

The scalation is smooth with 35–47 mid-body rows, 251–300 ventrals, single anal and 38–57 mainly divided subcaudals.

The Children's Python is one of the most common snakes in areas where it occurs. It is found during the day sheltering in rock crevices in rocky areas, where it is most common, and in hollow logs and other ground cover else-

317 Children's Python *Bothrochilus childreni* (Arnhem Hwy, NT).

318 Children's Python *Bothrochilus childreni*, unmarked specimen (Kunnanurra, WA).

where. This ground-dwelling python only rarely climbs trees. At night specimens are commonly seen crossing roads. This species feeds on small vertebrates, excluding frogs and fish.

It lays about twelve eggs around December. These hatch some two to four months later. Hatchlings measure about 25 cm.

319 Children's Python *Bothrochilus childreni*, unmarked specimen, head (Kunnanurra, WA).

Children's Python *Bothrochilus childreni*

WATER PYTHON
Bothrochilus fuscus Peters, 1873
Fig. 320

Sometimes measuring 3 metres in length, this snake averages about 2 metres. It is found throughout tropical Australia, usually near water.

The scalation is smooth with 41–55 mid-body rows, 270–300 ventrals, single anal and 60–90 paired subcaudals.

The Water Python is most common in swampy localities, and is particularly common around the large grassy swamps at the top end of the Northern Territory. This noctural snake feeds on a variety of vertebrates, and is a major predator of young crocodiles.

About ten eggs are laid at the beginning of the wet season. These hatch about eighty days later, with the young measuring about 45 cm.

320 Water Python *Bothrochilus fuscus* (Humpty Doo, NT).

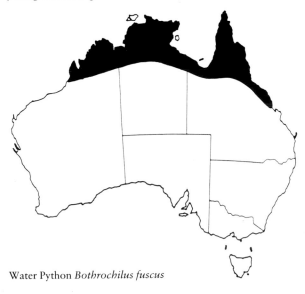

Water Python *Bothrochilus fuscus*

SPOTTED PYTHON
Bothrochilus maculosus Smith, 1985
Fig. 321

The Spotted Python is closely related to the Children's Python *Bothrochilus childreni*. The Spotted Python *Bothrochilus maculosus* is distinguishable by its different colour pattern and head morphology. Specimens up to 2 metres are known, but the average length is closer to about 80 cm. This species is known from coastal Queensland and nearby areas.

The scalation is smooth with 35–45 mid-body rows, 240–295 ventrals, single anal and 37–48 mainly divided subcaudals.

This ground-dwelling snake is almost invariably associated with rock outcrops, in which it lives, often in large numbers. A cave at Mount Etna, near Rockhampton, houses many thousands of Bent-winged Bats, which in turn supports a population of several hundred Spotted Pythons *Brothrochilus maculosus*, which feed on the bats.

127

321   Spotted Python *Bothrochilus maculosus* (Port Douglas, QLD).

322   Olive Python *Bothrochilus olivaceus* (Pine Creek, NT).

323   Olive Python *Bothrochilus olivaceus*, head (Pine Creek, NT).

The snakes congregate around the cave in these abnormally large numbers because of the apparently limitless food supply. Elsewhere this species feeds on a variety of vertebrates.

About ten eggs are produced around December which hatch about seventy to ninety days later. Hatchlings measure about 25 cm.

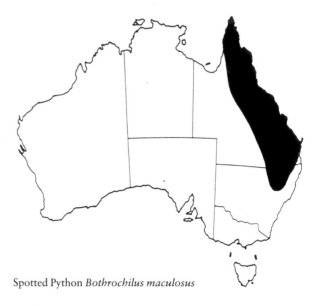

Spotted Python *Bothrochilus maculosus*

OLIVE PYTHON
Bothrochilus olivaceus Gray, 1842
Figs 322–324, 533

This snake is commonly confused with the Water Python *Bothrochilus fuscus*, but is distinguishable by its often larger size, different head shape and scalation. The Olive Python *Bothrochilus olivaceus* attains up to 3 metres or more in length. It is found throughout tropical and nearby parts of northern Australia, except east of Cape York. Largest specimens come from the Pilbara region in Western Australia.

The scalation is smooth with 55–80 mid-body rows, 340–415 ventrals, single anal and 90–110 divided subcaudals.

Olive Pythons are found in a variety of habitats, but

Olive Python *Bothrochilus olivaceus*

are most common in rocky hills, particularly near water. This mainly nocturnal snake feeds on smaller vertebrates and large specimens have been known to eat fully-grown kangaroos.

About ten eggs are produced, which take about seventy days to hatch, with hatchlings measuring about 37 cm.

ANT-HILL PYTHON
Bothrochilus perthensis Stull, 1932
Figs 25, 325–328, 537–540

This species was unknown for many years until I rediscovered it recently in the Pilbara region of Western Australia. This small python averages about 50–60 cm in length. The patterning on young specimens fades with age. In captive specimens the patterning often fades completely.

The scalation is smooth with 31–35 mid-body rows, 212–250 ventrals, single anal and 34–45 mainly divided subcaudals.

It is called the Ant-hill Python because of its habit of occupying the large termite mounds found where this snake occurs. The Ant-hill Python feeds principally on marsupial mice *Antechinus* sp., geckoes and skinks, all of which also live in the termite mounds. When termite mounds are not present this species can be found hiding in rock outcrops or spinifex bushes during the day. It is a strictly nocturnal species.

Specimens are always docile, even when freshly caught.

Mating occurs in winter, with about two or more very large eggs being produced in November–December which hatch sixty days later, around about January–February. Hatchlings measure about 20 cm.

To date, I am the only person to have successfully kept and bred this species. The breeding programme was terminated in July 1984 when NPWS officers stole all the snakes in order to fulfil a 'contract request'.

Ant-hill Python *Bothrochilus perthensis*

STIMSON'S PYTHON
Bothrochilus stimsoni Smith, 1985
Figs 28, 329

This snake is distinguishable from other small Python species by its well-defined pattern and distinctive elongat-

324 Olive Python *Bothrochilus olivaceus*, juvenile (Mount Isa, QLD).

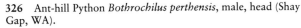

325 Ant-hill Python *Bothrochilus perthensis*, male (Shay Gap, WA).

326 Ant-hill Python *Bothrochilus perthensis*, male, head (Shay Gap, WA).

327 Ant-hill Python *Bothrochilus perthensis*, captive bred hatchling. Parents from Shay Gap, WA.

328 Ant-hill Python *Bothrochilus perthensis*(?) female (Katherine, NT).

ed head. It is found throughout most parts of Australia, except for the far north, east and south coastal regions.

The scalation is smooth with 39–47 mid-body rows, 260–302 ventrals, single anal and 40–53 mainly divided subcaudals.

This snake is commonly aggressive in temperament. Captive specimens often fail to settle down, even after several years. In the wild specimens are found in all types of habitat but mostly in rocky areas, although in the Pilbara and parts of the Northern Territory large numbers occupy large termite mounds.

This mainly nocturnal snake feeds on vertebrates. About ten eggs are laid in spring or early summer and hatch about seventy days later. Hatchlings measure 25 cm.

329 Stimson's Python *Bothrochilus stimsoni*, female (The Tits, WA).

330 Green Python *Chondropython viridis* (PNG).

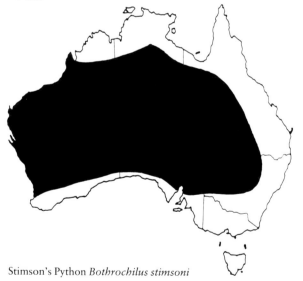

Stimson's Python *Bothrochilus stimsoni*

GREEN PYTHON
Chondropython viridis (Schlegel, 1872)
Figs 330, 331, 534

This snake attains 2 m. Common in New Guinea, it is also found in the Black Mountain area of far north Queensland, where it only occurs in very moist forest habitats. In Australia this snake is not common. Although adults are usually bright emerald green, some greeny-blue or blue

specimens are known. The adults that become bluish in colour, change to this colour from the 'usual' green form. Juveniles are either yellow or red in colour (red is dominant in heterozygous individuals). Colour change from red or yellow in juveniles to the adult green is both age and size related. The change takes place at between six months and three years of age, when the snakes are around 60 cm in length.

The scalation is smooth with 50–75 mid-body rows, 225–260 ventrals, single anal and 90–110 single subcaudals.

This snake is arboreal, and by day may be found resting on tree branches, in hollows, etc. It feeds mainly on birds but will also eat other animals. This snake adopts an unusual resting position in horizontal tree branches. It will coil into a series of folded loops, which are neatly proportioned so that the head is always resting in the centre. Only a similar South American boa is known to do this too.

Juveniles are known to use their tail as a lure to attract prey.

Some males of this species are known to have enlarged pelvic spurs, up to 3 cm long. These are used to hold the female's tail region in place and provide added stimulation when copulating.

This python lays about ten to sixteen eggs which hatch about fifty days later when incubated at 27 degrees celcius.

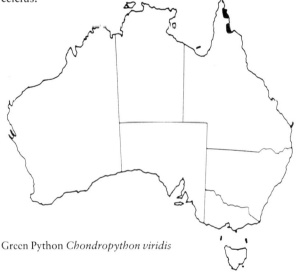

Green Python *Chondropython viridis*

SCRUB OR AMETHYSTINE PYTHON
Morelia amethistina (Schneider, 1801)
Figs 332, 333

Australia's largest python, this snake occurs from Mt Speck, near Townsville, North Queensland, to the tip of Cape York Peninsula. Specimens more than 5 metres long are known, although a 4-metre specimen is regarded as being large. Worrell (1970) mentions one specimen of 8.5 metres as being recorded from Greenhill near Cairns.

The scalation is smooth with 35–50 mid-body rows, 270–340 ventrals, single anal and 80–120 mainly divided subcaudals.

The Scrub Python is most common in rainforests and adjacent coastal habitats. This snake is commonly of uneven temperament and does not make a good pet. It is mainly diurnal in winter and nocturnal in summer, and feeds on a variety of large vertebrates, including fowls and wallabies. A four metre captive adult male maintained perfect health over several years by eating an average of

331 Green Python *Chondropython viridis*, in characteristic resting position (PNG).

332 Scrub Python *Morelia amethistina* (Hartley's Creek, QLD).

333 Scrub Python *Morelia amethistina*, head (Hartley's Creek, QLD).

334 Oenpelli Python *Morelia oenpelliensis* (Jabiru, NT).

335 Oenpelli Python *Morelia oenpelliensis*, head (Jabiru, NT).

nine plucked, size 14, supermarket chickens per year over several years. That indicated the relatively slow metabolic rate of this (and other) very large reptiles.

About nine to twenty eggs are laid before the wet season which hatch about seventy to ninety days later. The hatchlings measure about 65 cm.

Scrub Python *Morelia amethistina*

132

OENPELLI PYTHON
Morelia oenpelliensis (Gow, 1977)
Figs 334, 335

Found only in the rocky habitats and adjacent floodplains of Arnhem land in the Northern Territory, this snake averages between 3 and 4 metres in length. Unlike the Scrub Python *Morelia amethistina*, the Oenpelli Python *Morelia oenpelliensis* is a very docile snake. Although fairly common where it occurs, this snake was not discovered by white man until the 1970s. It has been suggested that this snake is very closely related to the Timor Python *Morelia* (?) *timorensis*.

The scalation is smooth, with 64–70 mid-body rows, 429–445 ventrals, single anal and 155–163 mainly divided subcaudals.

Specimens of this snake are known to have the ability to change colour pattern intensity, ranging from plain in colour to well-patterned within a few hours.

The Oenpelli Python is both diurnal and nocturnal and feeds on a variety of vertebrates.

The breeding biology of this species is not known.

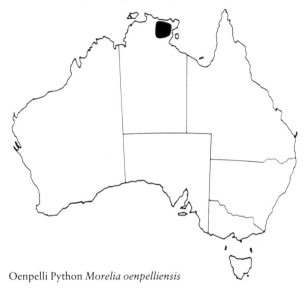

Oenpelli Python *Morelia oenpelliensis*

CARPET PYTHONS
Morelia spilota (Lacepede, 1804)
Figs 16, 24, 336–357

The taxonomic status of this species or species complex is one of the subjects most argued about by Australian snake experts. There are currently some six forms of Carpet Python recognised. Another closely related species, *Morelia carinata*, is known only from a single dead specimen held in the Perth museum. Because Carpet Pythons are common and well-known throughout Australia I have dealt with them in some detail here. The six forms of Carpet Python currently recognised are:

1 Diamond Python *Morelia spilota spilota*, found along the NSW coast south of Port Macquarie, and the near ranges south of Newcastle.

2 Carpet Python *Morelia spilota macropsila*, found along the north coast of New South Wales, and coastal Queensland and nearby areas.

3 Carpet Python *Morelia spilota* sub sp., found along the Murray-Darling river system of Victoria, South Australia, New South Wales and Queensland, and nearby areas.

4 Carpet Python *Morelia spilota variegata*, found along the coast and nearby areas of the Gulf of Carpentaria, the Northen Territory and north-west Western Australia.

5 Carpet Python *Morelia spilota bredli*, found around the Macdonnel Ranges and nearby parts of the inland of the Northern Territory.

6 Carpet Python *Morelia spilota imbricata*, found in the south-west corner of Western Australia.

Where the ranges of the different forms converge, intergrade specimens are found, although *Morelia spilota bredli* and *Morelia spilota imbricata* are apparently isolated populations. Photos of most forms of the Diamond or Carpet Python *Morelia spilota* are shown.

General. Carpet Pythons are found in most parts of mainland Australia, and the colour varies strongly with locality and within a given locality. In some areas juveniles are a different colour from adults. The average adult size of Carpet Pythons also varies with locality, with adult specimens averaging 2 metres in length occurring in north Queensland, the tropical north of Australia and inland eastern Australia and South Australia. Elsewhere specimens up to and over 3 metres occur. In south-east Queensland specimens of nearly 4 metres are known.

The temperament of specimens varies both with locality and even within a locality. Queensland, Northern Territory and Kimberley specimens tend to be unpredictable, while those from New South Wales, southern Western Australia, and inland areas tend to be docile.

This snake is diurnal in cold weather and nocturnal in warm weather. Carpet Pythons occupy all habitats, although in arid areas are usually found adjacent to water courses or rocky hills. Carpet Pythons are a climbing snake, and feed mainly on birds and mammals. This species commonly enters aviaries, where it feeds on birds and then is unable to escape through the wire because of the bulge in its body caused by the swallowed birds.

In certain periods Carpet Pythons are known to aggregate in large numbers, presumably for breeding purposes, and are among the few Australian snakes that are known to do this.

These snakes lay about twelve eggs in spring which hatch about seventy days later. One specimen is recorded as having produced forty-seven eggs. When coiled around the eggs, the female Carpet Python will 'shiver' and increase her own body temperature and that of the eggs above the ambient environmental temperature, to aid in the incubation of the eggs. Hatchling Carpet Pythons measure about 30 cm.

336   Carpet Python/NSW Diamond Python form *Morelia spilota spilota*, sub-adult, male (Kenthurst, NSW).

337   Carpet Python/Murray Valley form *Morelia spilota* sub sp., female (Carinda, NSW).

338   Carpet Python/Murray Valley form *Morelia spilota* sub sp., (Birdsville, QLD).

339   Carpet Python/Queensland form *Morelia spilota macropsila*, male (Bundaberg, QLD).

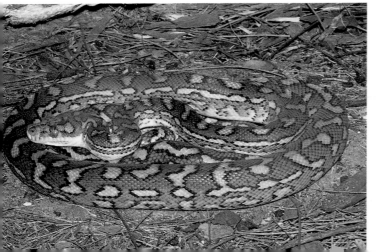

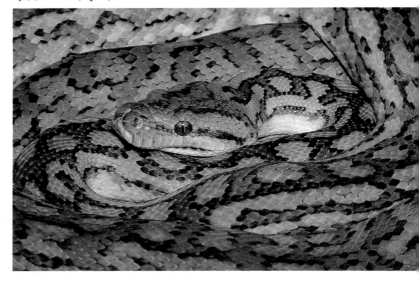

340   Carpet Python/Queensland form *Morelia spilota macropsila*, female, head (Bundaberg, QLD).

341   Carpet Python/Queensland form *Morelia spilota macropsila*, male (Townsville, QLD).

342   Carpet Python/Queensland form *Morelia spilota macropsila*, male (Townsville, QLD).

343   Carpet Python/Queensland form *Morelia spilota macropsila* (Coffs Harbour, NSW).

344   Carpet Python/Queensland form *Morelia spilota macropsila* (Port Douglas, QLD).

345   Carpet Python/Queensland form *Morelia spilota macropsila*, head (Port Douglas, QLD).

Carpet Pythons *Morelia Spilota*
(All Subspecies)

346 Carpet Python/Queensland form *Morelia spilota macropsila*, juvenile (locality unknown).

347 Carpet Python/Queensland form *Morelia spilota macropsila*, one-year-old juvenile, brown colour phase which changes with age (Bundaberg, QLD).

348 Carpet Python/Queensland form *Morelia spilota macropsila* (Grafton, NSW).

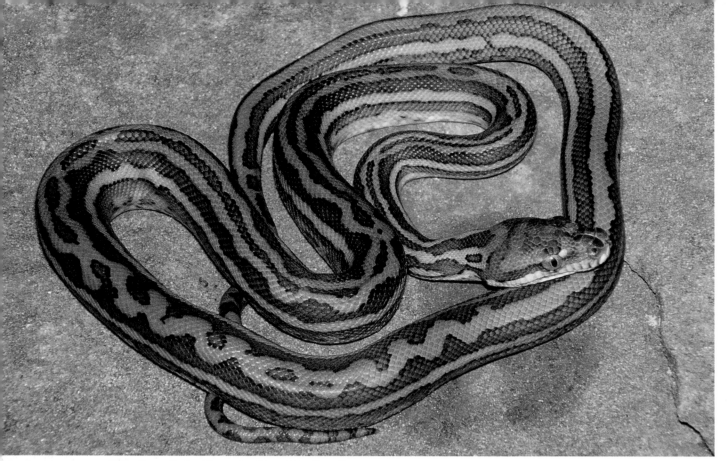

349 Carpet Python/Queensland form *Morelia spilota macropsila* (Cape York, QLD).

350 Carpet Python/Northern Territory form *Morelia spilota variegata* (Jabiru, NT).

352 Carpet Python/Northern Territory form *Morelia spilota variegata*, juvenile (Darwin, NT).

351 Carpet Python/Northern Territory form *Morelia spilota variegata*, head (Jabiru, NT).

353 Centralian Carpet Python *Morelia spilota bredli* (Alice Springs, NT).

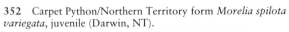

354 Carpet Python/hybrid between *M.s. spilota* and *M.s.* sub sp. (Murray Valley) (Bathurst, NSW).

355 Carpet Python/hybrid between *M.s. spilota* and *M.s.* sub sp. (Murray Valley), head (Bathurst, NSW).

356 Carpet Python/hybrid between *M.s. spilota* and *M.s. macropsila* (Barrington Tops, NSW).

357 Carpet Python/hybrid between *M.s. spilota* and *M.s. macropsila*, head (Barrington Tops, NSW).

PROBLEMS OF PYTHON CLASSIFICATION AND HYBRID PYTHONS

Figs. 358–361

With the recent exception of the Black-headed Python and Woma (Genus *Aspidites*), all other Australian Pythons have at various times been assigned to a number of different genera. Numerous schemes of classification for the remaining Australian species of Python have been proposed. These range from the placing of all species in the genus Python shared with other non-Australian species, to placing the species in question in up to six genera (namely *Bothrochilus*, *Chondropython*, *Liasis*, *Lisalia*, *Morelia* and *Python*). The assignment of given species within a particular genus is also a matter of conflict. For example, within the last ten years the Scrub Python *Morelia amethistina* has been placed in the genera *Liasis*, then *Python* and now *Morelia*.

A further problem is that some previously used generic names, including *Liasis*, are now thought not to be valid, and the issue as to which generic names should be used is not yet fully resolved.

In reality all Australian python species excluding *Aspidites* are fairly closely related, and should perhaps be placed in a single genus with further placement in subgenera. The conflict here is one between 'lumpers' who would agree with the above statement, and 'splitters' who would fear that by placing the Pythons in question into a single genus, the relationships between species might be obscured.

In the late 1970s the Royal Melbourne Zoo had a male Carpet Python *Morelia spilota* successfully breed with a female Scrub Python *Morelia amethistina* and a female Water Python *Bothrochilus fuscus*. The offspring produced were intermediate in characteristics between the parent snakes, and themselves appear to be fertile, although at the time of writing had not bred. Such an event indicates that the three species are closely related, and should in all probability be placed in a single genus. This indicates potential problems for the Darwinian classification of species.

Hybridisation and creation of 'new' species are two practices which conservationists generally condemn, for a number of reasons. However, the case cited above was probably of great benefit to Australian herpetology, and in the long term will probably assist in the conservation of Australian Pythons.

137

358 Hybrid between Carpet python *Morelia spilota* and Scrub Python *Morelia amethistina*. Snake is poised to strike. Note that the coloration is more typical of *Morelia spilota* than of *Morelia amethistina*.

359 Hybrid between Carpet Python *Morelia spilota* and Scrub Python *Morelia amethistina*, head. Note the missing scale at front of mouth, a congenital defect. Also note the large dorsal head shields typical of *Morelia amethistina*.

360 Hybrid between Carpet Python *Morelia spilota* and Water Python *Bothrochilus fuscus*. Note that the coloration is intermediate between the two species, and not typical of either.

361 Hybrid between Carpet Python *Morelia spilota* and Water Python *Bothrochilus fuscus*. Note the large dorsal head shields more typical of *Bothrochilus fuscus*.

139

Family Acrochordidae (File Snakes)

File snakes occur in the area from south-east Asia to northern Australia, with the two species occurring throughout this range.

These snakes are aquatic fish-feeders. They are very sluggish movers over dry land and have a very flabby appearance on land.

They are called File Snakes because of their coarse, finely pointed scales which give them a file-like texture. These scales encircle the body, so that there are no well-defined enlarged ventral scales.

File Snakes are harmless, solid-toothed snakes, although a bite from a large specimen can inflict severe wounds. These snakes will only bite, however, if very roughly handled. They have a prehensile tail.

File Snakes are live-bearing.

362   Arafura File Snake *Acrochordus arafurae*, male (Alligator River, NT).

363   Arafura File Snake *Acrochordus arafurae*, albino female (Alligator River, NT).

ARAFURA FILE SNAKE
Acrochordus arafurae McDowell, 1979
Figs 362–364

The Arafura File Snake grows to about 2 metres with females being by far the larger sex. It is found in permanent watercourses of northern Australia, from the west of Cape York, Queensland, to the north Kimberleys, Western Australia. A similar species, the Little File Snake *Acrochordus granulatus*, is found in estuarine habitats in northern Australia and south-east Asia.

The Arafura File Snake is nocturnal, and is most abundant in large lowland river systems with adjoining swamps and billabongs. Migrating specimens will freely enter salt water and even cross open seas. When swimming along the water surface, this snake makes a peculiar 'plop-plop-plop' by opening and closing its mouth. This is done by 'bubbling' the air and water.

This species is capable of sperm storage for several years.

From twenty to thirty live young are produced.

Arafura File Snake *Acrochordus arafurae*

Family Colubridae (Colubrid Snakes)

Colubrids are found only in tropical parts of Australia and along the east coast. About ten species occur within Australia. Those species are usually found elsewhere, and the invasion of this family into Australia occurred in recent geological history. Elsewhere in the world, Colubrids are the dominant type of snake.

Colubrids are an extremely diverse group of snakes in terms of physical forms, biology, etc. Species may be non-venomous or venomous. Venomous species inject their venom by fangs mounted at the rear of their mouth. However, no Australian Colubrid is regarded as being dangerous, the bites merely being of nuisance value.

KEELBACK OR FRESHWATER SNAKE
Amphiesma mairii (Gray, 1841)
Fig. 365

Common along the coasts of Queensland, the Northern Territory, the Kimberleys in Western Australia and

364 Arafura File Snake *Acrochordus arafurae*, albino female, head (Alligator River, NT).

nearby areas, the Keelback varies in colour from grey, green, red, olive, black or something between these colours. Often more than one colour phase occurs within a single locality.

Averaging slightly less than a metre in length, this harmless snake is commonly confused with the dangerous Rough-scaled Snake *Tropedichis carinatus*.

The scalation of the Keelback is strongly keeled, with 15–17 mid-body rows, 130–165 ventrals, divided anal and 50–85 divided subcaudals.

This snake is usually found close to water, where it feeds exclusively on frogs and the introduced Cane Toad *Bufo marinus*. This snake appears to have immunity to the venom of the Cane Toad, unlike most other reptiles. When feeding, this snake usually swallows frogs and toads hind quarters first.

The Keelback is both diurnal and nocturnal. When specimens are caught, they will often try to shake themselves loose, even by shedding their tail, which will not regenerate.

Mating occurs in winter and spring with six to twelve eggs being laid in January and February. These hatch about seventy days later. Hatchlings measure 14 cm.

365 Keelback *Amphiesma mairii* (Bundaberg, QLD).

Keelback *Amphiesma mairii*

366 Brown Tree Snake *Boiga irregularis*, banded northern form (Turkey Creek, WA).

367 Brown Tree Snake *Boiga irregularis* eastern and southern form (St Ives, NSW).

368 Brown Tree Snake *Boiga irregularis* eastern and southern form, head (West Head, NSW).

BROWN TREE SNAKE
Boiga irregularis (Merrem, 1802)
Figs 15, 366–368

The Brown Tree Snake is found along the east coast north of Sydney Harbour and in tropical parts of northern Australia. Specimens sometimes attain 2 metres in length. Two basic colour phases occur. A brick-red phase with some black markings is found east of Cape York and along the east coast. A red and white banded form occurs in other parts of tropical Australia. On Cape York itself these forms hybridise.

The scalation is smooth with 19–23 mid-body rows, 225–265 ventrals, single anal and 85–130 divided subcaudals.

This rear-fanged venomous snake has such a feeble venom apparatus that venom is rarely injected when it bites humans. When caught it is usually aggressive, striking repeatedly from the S-shaped coils in the forepart of its body.

The Brown Tree Snake is strictly nocturnal, and by day is usually found sheltering in rock outcrops or tree hollows. Around Sydney many specimens can be found in the sandstone honeycombs of caves in hilly areas. Colonies of up to six specimens are common. At night specimens can be seen crossing roads.

The Brown Tree Snake feeds on a variety of vertebrates, which are constricted strongly as well as being bitten in order to be killed.

This species lays about ten eggs, with hatchlings measuring about 28 cm in length.

Brown Tree Snake *Boiga irregularis*

GREEN TREE SNAKE
Dendrelaphis punctulatus (Gray, 1827)
Figs 369–371

Found along the coasts and nearby areas of northern and eastern Australia, the Green Tree Snake may attain up to 2 metres in length. The colour varies geographically from greys to greens, blues, browns, black or yellows. Usually the skin between the scales is green.

The scalation is smooth with 13–15 mid-body rows, 180–230 ventrals, divided anal and 100–150 divided subcaudals.

The harmless Green Tree Snake will inflate the forepart of the body and hiss when threatened, revealing the blue skin between the scales. It will also sometimes release a strong-smelling odour from anal glands. It is strictly diurnal and forages on the ground and in vegetation, feeding on small vertebrates, particularly frogs. It is a very fast-moving species.

When inactive the Green Tree Snake is usually found under ground cover, in tree hollows or rock crevices. Winter aggregations of up to six individuals are quite common, and this species will aggregate with other snake species.

About eight eggs are laid in mid summer, which hatch some sixty to eighty days later. Hatchlings measure 30 cm.

Green Tree Snake *Dendrelaphis punctulatus*

SLATEY-GREY SNAKE
Stegonotus cucullatus (Dumeril, Bibron and Dumeril, 1854)
Figs 372, 373

This snake ranges from brown to grey or black in colour, and attains 1.4 m. It is found in far north Queensland, tropi-

369 Green Tree Snake *Dendrelaphis punctulatus* (St Ives, NSW).

370 Green Tree Snake *Dendrelaphis punctulatus*, head (Hornsby, NSW).

371 Green Tree Snake *Dendrelaphis punctulatus*, black colour phase (Townsville, QLD).

372 Slatey-grey Snake *Stegonotus cucullatus* (Port Douglas, QLD).

373 Slatey-grey Snake *Stegonotus cucullatus*, head (Port Douglas, QLD).

cal parts of the Northern Territory and adjacent coastal areas.

The scalation is smooth with 17–19 mid-body rows, 170–225 ventrals, single anal and 65–105 divided subcaudals.

This harmless snake is usually aggressive when caught, biting repeatedly and exuding a strong odour from anal glands. In captivity, however, it settles down very rapidly to become a docile, easily handled snake.

The Slatey-grey Snake is nocturnal and although found in a variety of habitats, it is usually found in association with water. It feeds mainly on frogs.

An egg layer, it lays about twelve eggs per clutch. Hatchlings measure about 30 cm.

Slatey-Grey Snake *Stegonotus cucullatus*

Family Elapidae (Front-fanged Venomous Land Snakes)

Found in many parts of the world, the Elapids are the dominant snakes in most parts of Australia and Tasmania. (The only Tasmanian snakes are Elapids.) Elapids include some of the world's deadliest snakes, including the American Coral Snakes, African Mambas, Asian Kraits, Cobras from Afro-Asia, and all of Australia's dangerous land snakes.

Only in Australia has this family undergone a most dramatic adaptive radiation and speciation in the absence of the Colubrid and other snakes found elsewhere. About seventy Australian species are currently known. The larger species are all dangerous, with most smaller species being relatively harmless to humans, due to their small fangs and relatively weak venom.

Elapids are distinguishable by the fangs mounted at the front of their mouth. The fangs are shed and replaced at regular intervals. Venom is injected through them into prey by a hypodermic-like action. Elapid venom is most dangerous because of its neurotoxic (nerve-killing) effects, although other components present do cause complications in serious bites.

The biology and forms of different species are highly varied, with species occupying most available habitats. Reproductive modes range from egg-laying to live-bearing and most intermediate modes.

DEATH ADDER
Acanthophis antarcticus (Shaw and Nodder, 1802)
*Figs 1, 22, 23, 26, 29–33, **374–377**, 473, 480–506, 543–553, 558, 559*

This dangerous viper-like snake is found in most parts of New South Wales, the southern half of Queensland, and southern parts of Western Australia and South Australia. It might occur in parts of the Northern Territory.

The base colour is usually red or grey; however, within these parameters colour may range from yellows, reds, browns, greys, black or even green. Average adult length is 57 cm; however, specimens up to a metre long are known. Females generally attain a much larger size than males.

The scalation is slightly keeled with 21 mid-body rows, 110–130 ventrals, single anal and 40–55 subcaudals of which about half are divided.

Death Adders generally are not aggressive, but they will strike with great accuracy at anything coming within range, including airborne objects. When agitated it will flatten its entire body.

The Death Adder has the most effective biting mechanism of any Australian snake, and has highly neurotoxic venom. Before the development of antivenom about 60 per cent of all recorded Death Adder bites were fatal. This snake is particularly dangerous because of its habit of lying concealed in leaf litter on the ground, and not moving away when approached by humans. It then bites when trodden on.

Although this snake will occur in a variety of habitats, it is restricted to virgin bushland, where it is most common in low scrub habitats with plenty of leaf litter.

This species is mainly nocturnal, and is most active on warm nights with falling air pressure. Most specimens caught are mature males, looking for mates.

The food of this snake consists of all suitable vertebrates including birds, which it catches by wriggling its tail as a lure. The prey will come towards the tail thinking that it is some kind of worm or insect and be bitten while doing so.

Mating occurs at any time of year, although mainly in autumn and spring. Gestation appears to be from seven to nine months for this snake. From three to forty (average twenty) live young are born in late summer or early

374 Death Adder *Acanthophis antarcticus*, red form, female (West Head, NSW).

375 Death Adder *Acanthophis antarcticus*, grey form, female (West Head, NSW).

autumn. Most females only reproduce every second year, but some do so annually. This pattern of reproductive frequency is genetically determined and is not dependent on food availability for females. The young, which are more brightly coloured replicas of their parents, may be of red or grey base colour, even if from the same parent (again genetically determined, with red being the dominant allele/type of gene). They average 16–17 cm at birth, and take two or three years to become adults.

In captivity Death Adders have been bred more often than any other venomous Australian snake.

Death Adder *Acanthophis antarcticus*

376 Death Adder *Acanthophis antarcticus*, red form, male, head (West Head, NSW).

377 Death Adder *Acanthophis antarcticus*, female, South Australian coastal form (Port Pirrie, SA).

NORTHERN DEATH ADDER
Acanthophis praelongus Ramsay, 1877
Figs 378–380

Found in tropical and nearby parts of Australia, this dangerous snake is distinguished from the Death Adder *Acanthophis antarcticus* by its rugose head scales, usually raised above the eye, and different scalation. Like the

378   Northern Death Adder *Acanthophis praelongus*, male (Kunnanurra, WA).

379   Northern Death Adder *Acanthophis praelongus*, male, head (Kunnanurra, WA).

380   Northern Death Adders *Acanthophis praelongus*, females (red from Weipa, QLD, grey from Tully, QLD).

Death Adder, the colour is highly variable, but always derives from a red or grey base.

Average adult length is about 50 cm, with females generally the larger specimens.

The scalation is keeled or smooth, with 23 mid-body rows, 122–140 ventrals, single anal and 40–55 subcaudals of which about half are divided.

Like other Death Adders, this species is restricted to relatively virgin habitats. In wetter areas this snake is usually found in hilly and scrubby localities. In arid areas where this species occurs, it is restricted to areas of spinifex bush *Triodia* spp., where it takes refuge during the day.

The biology and habits of this snake are similar to that of the Death Adder. However, this snake is considerably more aggressive and, unlike the Death Adder, won't hesitate to bite when handled.

In North Queensland the populations of the Northern Death Adder have been decimated by the introduction of the Cane Toad *Bufo marinus*, which both feeds on small Northern Death Adders, and is poisonous to those which try to feed on the toads.

Young are born around February to March and average 14–15 cm at birth. Like other Death Adders, most females only reproduce every second year.

Northern Death Adder *Acanthophis praelongus*

DESERT DEATH ADDER
Acanthophis pyrrhus Boulenger, 1898
Figs 381–383, 554–557, 558

This dangerous type of Death Adder is restricted to desert parts of Australia. Most specimens are orangish in colour, although grey base colour snakes are known to exist. The highly rugose scalation distinguishes this snake from other Death Adders. The average adult length is about 50 cm, with females being the larger sex.

The scalation is keeled with 19–21 mid-body rows, 120–160 ventrals, single anal and 40–65 subcaudals of which about half are divided.

The Desert Death Adder is restricted to spinifex *Triodia* spp. country where it occurs. The removal of this grass by cattle graziers throughout most of arid Australia has led to a dramatic decline in numbers of this species, particularly in the Northern Territory. The Desert Death Adder is most common in the Pilbara region of Western Australia, where it still occurs in massive numbers. I have caught more than a hundred specimens per night in the Pilbara, when driving along roads. The sex ratio caught was one female per ten males, although my studies indicate that the sex ratio in the wild is 50 per cent males to 50 per cent females. (Males are obviously the more mobile sex.)

Desert Death Adders are more nervous in disposition than Death Adders *Acanthophis antarcticus* but are not aggressive. Desert Death Adders are very swift-moving and appear to stalk their prey more actively than other Death Adders. Desert Death Adders appear to have a very strong preference for lizards over other food types, and when kept in captivity it takes some time to get most specimens to feed regularly on mice. Also unlike other Death Adders, Desert Death Adders have strong cannibalistic tendencies.

About eight to twelve live young are born in mid to late summer. These measure about 15 cm at birth.

381 Desert Death Adder *Acanthophis pyrrhus*, male (Goldsworthy, WA).

Desert Death Adder *Acanthophis pyrrhus*

COPPERHEADS
Genus *Austrelaps* Worrell, 1963
Figs 385–389

These dangerous snakes are restricted to colder parts of south-eastern Australia and Tasmania. Usually attaining up to 1.8 metres, they are highly variable in colour, ranging from yellows, browns, reds, greys or black. Specimens may or may not have a different-coloured nape region.

Copperheads occur in three forms, recently reclassified as separate (but similar) species. They are:

1 Highland form (*Austrelaps ramsayi*), found in New South Wales and north-eastern Victoria.

2 Lowland form (*Austrelaps superbus*), found in Victoria, Tasmania and south-eastern South Australia.

3 Dwarf form (*Austrelaps labialis*), found in south-eastern South Australia.

(There is, however, some doubt at present as to whether these scientific names are appropriate.)

The scalation is smooth with 15 mid-body rows, 145–160 ventrals, single anal and 40–55 single subcaudals.

These heavily-built snakes are commonly confused with other venomous species. The venom is similar in constitution to that of the Tiger Snake *Notechis scutatus*, being neutralised by the same specific antivenom.

382 Desert Death Adder *Acanthophis pyrrhus*, female, immediately after sloughing (The Tits, WA).

383 Desert Death Adder *Acanthophis pyrrhus*, male, head (Goldsworthy, WA).

384 Barkly Tableland Death Adder *Acanthophis* sp. (possibly *Acanthophis antarcticus*) (Anthony's Lagoon, NT).

385 Highland Copperhead *Austrelaps ramsayi*, female (Wentworth Falls, NSW).

386 Highland Copperhead *Austrelaps ramsayi*, male (Tarana, NSW).

387 Highland Copperhead *Austrelaps ramsayi*, male, ventral surface (Tarana, NSW).

388 Highland Copperhead *Austrelaps ramsayi*, one-year-old juvenile (Zig-zag Railway, NSW).

389 Lowland Copperhead *Austrelaps superbus* (Mount Gambier, SA).

Copperheads are not aggressive and will always flee if given the opportunity.

Copperheads are most common near swamps and marshes where large numbers may occur within comparatively small areas. Copperheads spend most of their time concealed in vegetation and their presence is frequently undetected by both local residents and herpetologists.

The Copperhead feeds mainly on frogs which are abundant where they occur; however, it is an opportunistic feeder, eating all available, small enough vertebrates, including other snakes. Because of this, areas where Copperheads are common often lack other snake species.

In hot weather Copperheads are crepuscular or nocturnal, and otherwise are diurnal. The Copperhead has a stronger resistance to cold than other snakes, having shorter periods of winter dormancy than other snakes in the same areas.

Mating occurs in early spring with an average of fourteen live young being born in late summer. The young average 18 cm in length.

390 White-crowned Snake *Cacophis hariettae* (Port Douglas, QLD).

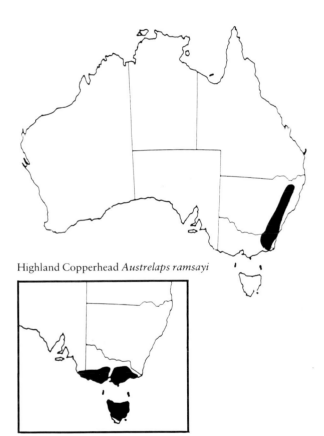

Highland Copperhead *Austrelaps ramsayi*

Lowland Copperhead *Austrelaps superbus*

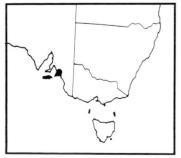

Dwarf Copperhead *Austrelaps labialis*

391 White-crowned Snake *Cacophis hariettae*, head (Port Douglas, QLD).

WHITE-CROWNED SNAKE
Cacophis hariettae Krefft, 1869
Figs 390, 391

This snake attains 40 cm and is found from far north Queensland, along the coast and adjacent ranges to the far north coast of New South Wales. Its dorsal colour is usually brownish, but will turn greyish before sloughing. Females are usually the larger sex.

The scalation is smooth with 15 mid-body rows, 170–200 ventrals, divided anal and 25–45 divided subcaudals.

Although found in a variety of habitats this snake is most common in forest habitats. By day this nocturnal snake shelters under ground cover. At night it feeds mainly on small lizards.

When caught, the White-crowned Snake will raise the forepart of its body and make repeated thrashing strikes with its mouth closed. It is rare to be bitten by this snake.

This species lays from two to ten eggs in summer, with hatchlings measuring 13 cm.

392   Krefft's Dwarf Snake *Cacophis krefftii* (Wyong, NSW).

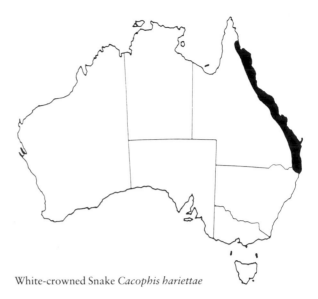

White-crowned Snake *Cacophis hariettae*

393   Krefft's Dwarf Snake *Cacophis krefftii*, ventral surface (Wyong, NSW).

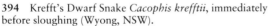

394   Krefft's Dwarf Snake *Cacophis krefftii*, immediately before sloughing (Wyong, NSW).

KREFFT'S DWARF SNAKE
Cacophis krefftii Gunther, 1863
Figs 392–394

This snake is found from Woy Woy on the New South Wales central coast to south-eastern Queensland. Its colour is greyish before sloughing. Males average 25 cm while females average 30 cm.

The scalation is smooth with 15 mid-body rows, 140–160 ventrals, divided anal and 25–40 divided subcaudals.

Found in forest habitats, this nocturnal snake feeds principally on small lizards. Most specimens can be found during the day hiding under ground cover such as rocks, logs, tree bark and thick leaf litter.

Like other snakes of the genus *Cacophis*, the Krefft's Dwarf Snake will bluff using a threat display, but rarely biting, when caught.

This snake lays an average of three large eggs. Hatchlings measure 10 cm.

Krefft's Dwarf Snake *Cacophis krefftii*

GOLDEN-CROWNED SNAKE
Cacophis squamulosus (Dumeril, Bibron and Dumeril, 1854)
Figs 27, 395, 396

This distinctive snake is found from southern New South Wales, along the coast and near ranges to central coastal Queensland. This snake averages 50 cm although females are the larger sex and large females are known to exceed 70 cm.

The scalation is smooth with 15 mid-body rows, 170–185 ventrals, divided anal and 30–52 divided subcaudals.

The Golden-crowned Snake is found in a variety of habitats, although when found in drier areas it is usually found close to water. Specimens caught during the day are found under a variety of cover including ground litter, well-embedded rocks, and even in hollow upright trees and bark on the sides of trees. At night this nocturnal species is commonly found crossing roads. The Golden-crowned Snake is a skink feeder and is often active on nights considered too cold for most other types of snake.

Golden-crowned Snakes are common in the inner suburbs of Sydney and Brisbane, and are commonly brought into houses by domestic cats. These snakes will adopt a fierce raised striking posture when caught, but rarely actually bite.

Mating occurs in spring with an average of six large eggs being produced in summer. Hatchlings measure 16 cm.

Golden-crowned Snake *Cacophis squamulosus*

SMALL-EYED SNAKE
Cryptophis nigrescens (Gunther, 1862)
Figs 397, 398

Found from Melbourne, Victoria, along the east coast, ranges and near slopes of Victoria, New South Wales and Queensland to the tip of Cape York, this dangerous snake is usually black, grey or dark brown in colour. Adults average 50 cm although some Queensland specimens, which tend to grow larger than southern specimens, may exceed a metre in length. A closely related species occurs in the coastal parts of the Northern Territory and adjacent parts of Western Australia.

The scalation is smooth with 15 mid-body rows, 165–210 ventrals, single anal and 30–45 single subcaudals.

395 Golden-crowned Snake *Cacophis squamulosus* (Terry Hills, NSW).

396 Golden-crowned Snake *Cacophis squamulosus*, head (Castlecove, NSW).

397 Small-eyed Snake *Cryptophis nigrescens* (Bundaberg, QLD).

151

398 Small-eyed Snake *Cryptophis nigrescens*, head (Port Douglas, QLD).

Most specimens of this nocturnal snake are caught during the day under cover, particularly in cold weather. When caught this snake will usually attempt to bite and may flatten its body while doing so. This snake has been found inside termite mounds, where it feeds on lizards that enter the mounds. The Small-eyed Snake is essentially a skink- and gecko-feeder.

Winter aggregations of this snake numbering up to nearly thirty individuals have been found around Sydney. It is believed that mating occurs in late autumn, winter and spring with two to eight (average five) live young being produced in late summer. Young measure 10–12 cm at birth.

Small-eyed Snake *Cryptophis nigrescens*

399 Black Whip Snake *Demansia atra* (Bundaberg, QLD).

BLACK WHIP SNAKE
Demansia atra (Macleay, 1884)
Fig. 399

Found along the coast and nearby areas of northern and north-eastern Australia, the Black Whip Snake attains about a metre in length, although Worrell (1970) records a specimen of nearly 2 metres. The colour ranges from light olive-brown to black.

400 Yellow-faced Whip Snake *Demansia psammophis* (Wyong, NSW).

This wholly nocturnal snake has killed at least two people from its bite, and although both deaths were the result of 'exceptional' circumstances this snake should be treated with care. Most bites, however, seem to result in little more than local swelling.

Black Whip Snake *Demansia atra*

The scalation is smooth with 15 mid-body rows, 160–220 ventrals, divided anal and 70–95 divided subcaudals.

The Black Whip Snake is one of Australia's fastest-moving snakes. Diurnal in habits, this species is most common in dry rocky and woodland habitats. When inactive this species hides under ground litter and is particularly fond of sheet metal. It feeds principally on lizards which it actively chases. Male combat is known to occur in this species.

From four to thirteen eggs are produced. Hatchlings measure about 16 cm.

YELLOW-FACED WHIP SNAKE
Demansia psammophis (Schlegel, 1873)
Fig. 400

The Yellow-faced Whip Snake is found in most parts of mainland Australia. Colourwise this snake is one of the most variable in Australia. Specimens range in colour from greys and greens to yellows, often with reddish markings running along the back. The scales may or may not form a reticulated pattern. Although the average length is 65 cm, specimens of more than a metre are known.

The scalation is smooth with 15 mid-body rows, 165–230 ventrals, single anal and 68–105 paired subcaudals.

The Yellow-faced Whip Snake is found in all types of habitat. This swift-moving diurnal species feeds principally on lizards which it actively chases. When inactive it hides under ground litter such as rocks, logs, bark, etc. During the colder months adult pairs are commonly found together in a single site, indicating that this is the mating season. Winter aggregations of up to five specimens are commonly found.

Males are usually the larger specimens and male combat, in the form of two males twisting around one another and biting each other, is known. Male combat occurs during the mating season, but in this species is not necessarily directly connected with the seduction of a mate.

From three to nine eggs are laid, often communally, in summer by this snake. The eggs take about eight to nine weeks to hatch and the hatchlings measure 15 cm.

When caught, this species will not hesitate to bite its captor, causing painful local swelling.

Yellow-faced Whip Snake
Demansia psammophis

COLLARED WHIP SNAKE
Demansia torquata (Gunther, 1862)
Fig. 401

Found throughout most of Queensland and adjacent parts of other states, the Collared Whip Snake attains 50 cm. More than one species is currently recognised here, but awaits formal taxonomic revision.

The scalation is smooth with 15 mid-body rows, 185–220 ventrals, divided anal and 70–90 divided subcaudals.

This fast-moving diurnal species is found in a variety of habitats ranging from arid grasslands to tropical forests. It feeds principally on skinks which it actively chases. When inactive, it is usually caught under loose ground cover, including sheets of tin, cardboard boxes, etc.

It lays from two to eight eggs.

Collared Whip Snake *Demansia torquata*

DE VIS' BANDED SNAKE
Denisonia devisi Waite and Longman, 1920
Figs 402, 403

Found throughout inland parts of Queensland and New South Wales, this snake varies in colour from brownish to whitish, depending on locality. It attains 50–60 cm in length. This species is commonly confused with the Tiger Snake *Notechis scutatus* and Death Adder *Acanthophis antarcticus*.

The scalation is smooth with 17 mid-body rows, 124–137 ventrals, single anal and 24–36 single subcaudals.

401 Collared Whip Snake *Demansia torquata* (Three Ways, NT).

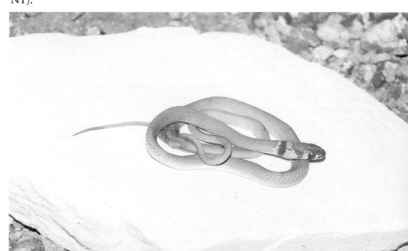

402 De Vis' Banded Snake *Denisonia devisi* (Bourke, NSW).

403 De Vis' Banded Snake *Denisonia devisi*, head (Moonie, QLD).

404 Rosen's Snake *Denisonia fasciata* (Shay Gap, WA).

When caught this snake will hold its body in stiff flattened curves and bite when the opportunity arises. It is a nocturnal species, although most specimens are found during the day sheltering under ground litter. The De Vis' Banded Snake is usually found near water, and is common along black-soil river flats throughout its range. It is principally a frog-feeder.

About eight live young are produced in summer.

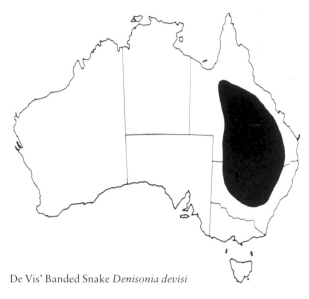

De Vis' Banded Snake *Denisonia devisi*

ROSEN'S SNAKE
Denisonia fasciata Rosen, 1905
Fig. 404

This snake is found in drier parts of Western Australia south of the Kimberley Ranges. It attains up to 60 cm in length.

The scalation is smooth with 17–19 mid-body rows, 140–185 ventrals, single anal and 20–40 single subcaudals.

This nocturnal snake is usually caught crossing roads at night where it is found. By day it may be found under ground cover or sheltering in Spinifex *Triodia* spp. bushes. It feeds principally on lizards.

A live-bearer, its breeding biology is little known.

405 Little Spotted Snake *Denisonia punctata* (Broome, WA).

Rosen's Snake *Denisonia fasciata*

LITTLE SPOTTED SNAKE
Denisonia punctata Boulenger, 1896
Fig. 405

Found from western Queensland throughout the top third of Australia, this snake attains 45 cm in length.

The scalation is smooth with 15 mid-body rows, 150–170 ventrals, single anal and 20–34 single subcaudals.

This nocturnal snake is usually found in low country and flood plains in a variety of habitats. Many specimens may be found within relatively small areas. It is principally a lizard-feeder. When caught it will adopt the defensive posture of flattening its body and striking frequently.

Two to five live young are produced.

Little Spotted Snake *Denisonia punctata*

WHITE-LIPPED SNAKE
Drysdalia coronoides (Gunther, 1858)
Fig. 406

The White-lipped Snake varies in colour from greys, greens, reds, browns or even black. It is found in colder parts of New South Wales, Victoria and all of Tasmania. It averages 40 cm in length.

The scalation is smooth with 15 mid-body rows, 120–160 ventrals, single anal and 35–70 single subcaudals.

This mainly diurnal snake is usually found sheltering under ground litter, including well-embedded rocks, when inactive. It is usually very abundant where it occurs. The White-lipped Snake feeds on skinks which it actively stalks.

About six live young are produced from January to early March and, unlike most kinds of snake, this species continues to feed throughout its pregnancy. The young measure about 10 cm at birth.

White-lipped Snake *Drysdalia coronoides*

EASTERN MASTERS SNAKE
Drysdalia rhodogaster (Jan, 1873)
Figs 407–409

Found along the New South Wales south coast and ranges, and in the Blue Mountains west of Sydney, this snake attains 50 cm.

406   White-lipped Snake *Drysdalia coronoides*, female (Lithgow, NSW).

The scalation is smooth with 15 mid-body rows, 140–160 ventrals, single anal and 40–55 single subcaudals.

This skink-feeding snake is usually active by day, and is mostly found either when active or sheltering under ground litter, particularly human rubbish. Where it occurs it is often found in large numbers.

Mating occurs in the cooler months, with about four live young produced during summer, measuring about 10 cm at birth. The young snakes are much darker in colour than their parents. In New South Wales the young are reproductive at three years old.

Eastern Masters Snake *Drysdalia rhodogaster*

RED-NAPED SNAKE
Furina diadema (Schlegel, 1837)
Fig. 410

Common throughout New South Wales, adjoining parts of South Australia and south-eastern Queensland, the Red-naped Snake grows to about 40 cm. The Red-naped Snake is commonly sent to museums on account of its unusual coloration.

407   Eastern Masters Snake *Drysdalia rhodogaster*, gravid female (Bendalong, NSW).

408   Eastern Masters Snake *Drysdalia rhodogaster*, male (Bendalong, NSW).

409   Eastern Masters Snake *Drysdalia rhodogaster*, head (Bendalong, NSW).

The scalation is smooth with 15 mid-body rows, 160–210 ventrals, divided anal and 35–70 divided subcaudals.

This snake is nocturnal in habit, although it is strongly crepuscular. When caught this snake will stiffen the forepart of its body and strike, but rarely actually bite.

The Red-naped Snake is found in all types of habitat. Specimens caught by day are found under ground litter, while at night they frequently cross roads. In forested areas this species is particularly common around disused rubbish tips, and several specimens may be found sharing the same site. The Red-naped Snake has been known to occupy the same site as the Yellow-faced Whip Snake *Demansia psammophis* for hibernation purposes.

Red-naped Snakes are commonly found inside termite mounds and for years it was believed that this species fed on termites. Captive observations and dissection of wild-caught specimens showed that this snake was essentially a skink-feeder.

Mating takes place in the cooler months with one to five eggs being produced in summer. Hatchlings measure about 8 cm in length.

410   Red-naped Snake *Furina diadema* (West Head, NSW).

Red-naped Snake *Furina diadema*

411   Moon Snake *Furina ornata* (Lake Argyle, WA).

MOON OR ORANGE-NAPED SNAKE
Furina ornata (Gray, 1842)
Fig. 411

Found in most parts of Australia where the Red-naped Snake does not occur, the Moon Snake is found throughout the north and western two-thirds of Australia. It is similar in appearance to the Red-naped Snake. It attains about 50 cm.

The scalation is smooth with 15–17 mid-body rows, 160–240 ventrals, divided anal and 35–70 divided subcaudals.

The biology of this snake is similar to that of the Red-naped Snake. The Moon Snake is found in all types of habitat, and most specimens caught are found crossing roads at night. It feeds principally on lizards.

One to five eggs are produced in summer and hatchlings measure about 8 cm.

Moon Snake *Furina ornata*

412   Swamp Snake *Hemiaspis signata* (Northbridge, NSW).

413   Swamp Snake *Hemiaspis signata*, head (Northbridge, NSW).

414   Pale-headed Snake *Hoplocephalus bitorquatus* (Moonie, QLD).

SWAMP OR MARSH SNAKE
Hemiaspis signata (Jan, 1859)
Figs *412, 413*

Found along the east coast of New South Wales and Queensland, the Swamp Snake is usually olive in colour although melanistic specimens from New South Wales are known. Although this species averages 50 cm in length, 90-cm females have been recorded from Engadine, just south of Sydney. Although most larger specimens I have caught have been females, detailed studies by Richard Shine indicate that there is little difference in adult sizes between the sexes, and that males may actually average slightly larger size than females.

The scalation is smooth with 17 mid-body rows, 153–170 ventrals, divided anal and 41–56 single subcaudals.

Swamp Snakes are usually found in marshy country, wet forests or adjacent to sand dunes. This species will aggregate in large numbers in areas of suitable habitat such as rubbish tips. This snake is mainly diurnal in habit and most specimens are caught during the day either on the move or under any suitable ground cover.

This snake feeds principally on frogs and skinks. Specimens caught and put in bags with other reptiles won't hesitate to eat them, even if the bags are being carried around.

Mating occurs in late autumn, winter and spring, with live young being produced in late summer. Although the average number of young produced is about six, larger females may produce up to twenty young. The young, which measure about 10 cm, are brightly coloured and have velvety black or dark heads.

Swamp Snake *Hemiaspis signata*

PALE-HEADED SNAKE
Hoplocephalus bitorquatus (Jan, 1859)
Figs *414, 415*

The potentially dangerous Pale-headed Snake is found along the coastal ranges slopes and adjacent plains of New South Wales from Ourimbah and Dubbo in the south, to the bottom of Cape York in North Queensland. It attains 50 cm although specimens of nearly a metre are known.

The scalation is smooth with 19–21 mid-body rows, 190–225 ventrals, single anal and 40–65 single subcaudals.

Although found in a variety of habitats, this snake is

most common in dry woodland habitats. It is a nocturnal tree-dweller and most specimens are caught either resting or moving on tree trunks.

Like all snakes of the genus *Hoplocephalus*, this snake is highly aggressive and when cornered will raise the forepart of its body into a series of S-shaped curves from which it will repeatedly strike at anything that comes within range.

Although this snake is not generally regarded as dangerous, children and elderly people may be at risk from its bite, and from other snakes of the genus *Hoplocephalus*.

The Pale-headed Snake feeds on a range of vertebrates. Mating is in the cooler months with two to eleven (average five) young being produced in mid summer.

The young measure 16 cm at birth.

Pale-headed Snake *Hoplocephalus bitorquatus*

BROAD-HEADED SNAKE
Hoplocephalus bungaroides (Schlegel, 1837)
Figs 416, 417

The Broad-headed Snake is restricted to the Hawkesbury sandstone formations found within a 250-km radius of Sydney, New South Wales. This 60-cm snake is commonly confused with the harmless Diamond Python *Morelia spilota spilota* found in the same areas. A serious bite from the very aggressive Broad-headed Snake may result in the need for antivenom (Tiger Snake *Notechis*) to be administered.

The scalation is smooth with 21 mid-body rows, 200–221 ventrals, single anal, and 40–60 single subcaudals.

This species lives under large exfoliating slabs of sandstone, crevices and similar places. Because of the immense habitat destruction occurring throughout its range this species is in serious decline. The stated policy of the New South Wales National Parks and Wildlife Service to prevent any captive breeding programmes seems to be ensuring this snake a certain path towards extinction. This snake readily breeds in captivity.

Most specimens of this snake are found in suitable isolated habitat, during autumn and spring. Broad-headed Snakes are mainly nocturnal although they are diurnal in mid winter, due to the fact that they are commonly found in cold places such as the ranges to the south and west of Sydney. Although opportunistic feeders, this species feeds mainly on Lesueur's Geckoes *Oedura lesueurii* in the wild.

Mating is in the cooler months, with five to twelve (usually about six) live young being produced in mid summer. The young measure 16 cm at birth.

Broad-headed Snake
Hoplocephalus bungaroides

415 Pale-headed Snake *Hoplocephalus bitorquatus*, head (Moonie, QLD).

416 Broad-headed Snake *Hoplocephalus bungaroides* (Lawson, NSW).

418 Stephen's Banded Snake *Hoplocephalus stephensi* (Mount Glorious, QLD).

STEPHEN'S BANDED SNAKE
Hoplocephalus stephensii Krefft, 1869
Figs 418, 419

The Stephen's Banded Snake is usually banded, although unbanded specimens are known to occur. I caught an unusually spotted specimen, without bands, at Mount Glorious, Queensland. This species is found in forest habitats north of Gosford, New South Wales, along the coast and ranges to south-eastern Queensland. Average length is 50 cm. Although not regarded as dangerous, this very aggressive snake should be handled with care.

The scalation is smooth with 21 mid-body rows, 220–250 ventrals, single anal and 50–70 single subcaudals.

417 Broad-headed Snake *Hoplocephalus bungaroides*, head (Lawson, NSW).

This aboreal snake is usually found at night crossing roads where it occurs. By day specimens are found under loose tree bark and in hollow tree limbs. It feeds on small lizards, frogs, small mammals and young birds.

Mating is in autumn, winter and early spring, with about six live young being produced in summer. The young measure 16 cm at birth.

Stephen's Banded Snake
Hoplocephalus stephensi

BLACK, WESTERN AND ISLAND TIGER SNAKES
Notechis ater (Krefft, 1866)
Figs 420–425

This deadly snake is found in Tasmania, offshore islands of Bass Strait and South Australia, isolated pockets of

419 Stephen's Banded Snake *Hoplocephalus stephensi*, head (Mount Glorious, QLD).

south-eastern South Australia, and south-western Western Australia. A number of distinct forms are recognised and about five have been formally classified as subspecies. Other forms await closer study and classification.

This snake is usually black, or very dark in colour, particularly specimens from south-eastern Australia. Western Australian specimens are more variable in colour. It typically attains 1.5 m in length, but average size varies strongly with locality. On some islands specimens regularly exceed 2 metres, whereas on adjacent islands maximums of half that length, and much lesser body weight, are the norm. Although all subspecies of this snake are deadly, venom yields and toxicity of specimens also varies strongly with locality.

The scalation is smooth with 15–21 mid-body rows, 155–190 ventrals, single anal and 40–60 single subcaudals.

This snake is usually more placid in nature than the Eastern Tiger Snake *Notechis scutatus*, and consequently bites are rare. A notable exception are specimens from the Broughton River area in South Australia, which tend to be very excitable and potentially aggressive.

Although found in a variety of habitats, this species is most common in rocky areas, marshlands and coastal dune grasslands. The Black Tiger Snake is nocturnal in hot weather, but otherwise is usually diurnal.

This species feeds on a variety of vertebrates, depending on location. Adults from some Bass Strait islands feed exclusively on young Mutton Birds, for about two months of the year, while apparently starving the rest of the year. King Island Tiger Snakes *Notechis ater humphreysi* are cannibalistic in nature.

About sixteen live young are produced in mid to late summer. These measure 15 cm, and if provided with plenty of food can attain maturity within two years.

This species is extremely hardy in captivity.

420 King Island/Tasmanian Tiger Snake *Notechis ater humphreysi* (King Island, Bass Strait).

421 King Island/Tasmanian Tiger Snake *Notechis ater humphreysi*, head (King Island, Bass Strait).

422 Chappell Island Tiger Snake *Notechis ater serventyi*, black-banded form (Chappell Island, Bass Strait).

Black, Western and Island Tiger Snakes *Notechis ater*

EASTERN TIGER SNAKE
Notechis scutatus (Peters, 1861)
Figs 17–20, 462–429, 542

The Eastern Tiger Snake is highly variable in colour, ranging through olives, yellows, browns, reddish, greys or even black, with or without cross bands of varying intensity. This 1.3-metre snake is highly dangerous, and has claimed the lives of more white Australians in the last 200 years than any other type of snake. It is found throughout

423   Chappell Island Tiger Snake *Notechis ater serventyi*, brown unbanded form (Chappell Island, Bass Strait).

424   Chappell Island Tiger Snake *Notechis ater serventyi*, head, brown unbanded form (Chappell Island, Bass Strait).

425   Broughton River Tiger Snake *Notechis ater* sub sp. (Broughton River, SA).

south-eastern Australia, except in very arid places, and areas where the Black Tiger Snake *Notechis ater* occurs.

The scalation is smooth with 15–19 mid-body rows, 146–185 ventrals, single anal and 39–65 single subcaudals.

The Eastern Tiger Snake is only aggressive when agitated, and when aroused will flatten its neck and body, and lunge forward in various directions, striking when possible. Although the Eastern Tiger Snake will occupy a variety of habitats, in hotter and drier places it tends to be found in the vicinity of water or on floodplains. It is most common in sandy and swampy locations.

Although usually diurnal, the Eastern Tiger Snake may become nocturnal in warm weather, particularly juveniles. Most specimens are caught during the day, either foraging or under loose ground cover. Diet is varied, but includes most types of small vertebrate, particularly frogs. The Cane Toad *Bufo marinus* is exterminating this snake in Queensland and parts of New South Wales.

Mating occurs in early spring with 17–109 (average 35) live young being produced in late summer. Young when born measure about 18 cm, and often have distinct thick, brown and yellow bands which fade in later life.

Eastern Tiger Snake *Notechis scutatus*

INLAND TAIPAN
Oxyuranus microlepidotus (McCoy, 1879)
Fig. 430

This dangerous snake attains 2 metres in length, although specimens of nearly 3 metres are known. The colour of this snake is brownish, but changes seasonally. In winter specimens are dark brown with blackish heads while in summer specimens are light-brownish colour without a blackish head. This snake is found on black-soil plains of the channel country of the Northern Territory, Queensland, South Australia and north-western New South Wales. Old records from last century show this snake as occurring along the Murray–Darling river system but none have been found in this area since then.

The scalation is smooth with 23 mid-body rows, 220–250 ventrals, anal is usually single, and 55–70 divided subcaudals.

This snake shares with the Mainland Taipan *Oxyuranus scutellatus* the distinction of being Australia's deadliest snake, in terms of venom deadliness per bite. An average bite can kill 50,000 mice — about fifty human lethal doses in laboratory terms.

426 Eastern Tiger Snake *Notechis scutatus*, obese captive specimen (Lake George, NSW).

427 Eastern Tiger Snake *Notechis scutatus*, head (Lake George, NSW).

428 Eastern Tiger Snake *Notechis scutatus*, one-year-old juvenile (Tarana, NSW).

429 Eastern Tiger Snake *Notechis scutatus*, two-year-old specimen (West Head, NSW).

430 Inland Taipan *Oxyuranus microlepidotus* (Birdsville, QLD).

431 Mainland Taipan *Oxyuranus scutellatus* (Cairns, QLD).

432 Mainland Taipan *Oxyuranus scutellatus*, head (Jellaten, QLD).

433 King Brown Snake *Pseudechis australis* (Winton, QLD).

The Inland Taipan is diurnal and lives on black-soil plains where it occurs. When inactive it hides in the large cracks that form in this soil. This snake (which is often common where it occurs) is mostly caught when foraging over the plains in the morning and at dusk, when it is not too hot. Most specimens are seen in spring and autumn.

164

When this snake bites its prey it hangs on, unlike the Mainland Taipan *Oxyuranus scutellatus* which lets go of it. In the wild it is believed that the Inland Taipan feeds principally on the plague rat *Rattus villosissimus*. This snake is not at all aggressive, being more placid in nature than most other deadly snakes.

About twelve eggs are produced in summer. Hatchlings measure about 40 cm.

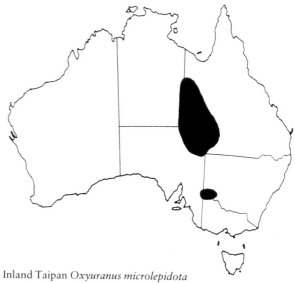

Inland Taipan *Oxyuranus microlepidota*

MAINLAND TAIPAN
Oxyuranus scutellatus (Peters, 1867)
Figs 431, 432

This deadly 2-metre snake is found along the coast and nearby areas of Queensland and far northern New South Wales. It is also found in tropical parts of the Northern Territory, Western Australia and parts of New Guinea.

The Mainland Taipan is recognisable by its coffin-shaped head. Large Taipans (all types) are, by popular definition, the deadliest land snakes on earth, and this is one of Australia's most dangerous snakes. Before the development of antivenom, hardly anyone survived a bite from this snake. Humans have been known to die within an hour from its bite.

The scalation is smooth with 21–23 mid-body rows, 220–250 ventrals, single anal and 45–80 divided subcaudals.

The fast-moving diurnal Mainland Taipan is common in a variety of habitats, and is abundant in the sugar-cane-growing areas of Queensland, where it thrives on introduced rats and mice. Herpetologists in North Queensland capture most specimens around wind rows, which are found in newly-cleared areas. This species is caught in large numbers around Julatten, north of Cairns, Queensland. The Mainland Taipan is, however, an opportunistic feeder, preying on a range of vertebrates. It will let go of its prey after biting it, waiting for its venom to kill the prey. It has very keen senses of smell and sight and is able to detect even the slightest movements.

Although a very shy species, the Mainland Taipan will stand its ground when cornered, and strike rapidly, often inflicting multiple bites.

Male combat occurs in this species.

Seven to twenty eggs are laid by this snake in summer. This species readily breeds in captivity. Hatchlings measure 40 cm.

Mainland Taipan *Oxyuranus scutellatus*

King Brown Snake *Pseudechis australis*

KING BROWN SNAKE
Pseudechis australis (Gray, 1842)
Figs 433–435

The deadly King Brown Snake is actually a close relative of the Black Snakes. It averages about 1.8 metres in length, although specimens of nearly 3 metres are known. Colour varies through various shades of reds, browns, yellow and even nearly black. This thickly-built snake is found throughout most parts of Australia except for the far south and south-eastern coastal regions. A bite from this snake yields a greater quantity of venom than that of any other Australian snake.

The scalation is smooth with 17 mid-body rows, 189–220 ventrals, divided anal and 50–75 subcaudals of which about half are divided.

The King Brown Snake occurs in all types of habitat but is most common in dry woodland and arid habitats. It is diurnal in cold weather and both diurnal and nocturnal in warm weather.

The King Brown Snake is not highly aggressive, preferring to flee rather than bite when disturbed. This snake may flatten its neck when harassed. It feeds on a variety of vertebrates including other snakes. The venom of this snake seems to have a potent effect on other snakes, while the venom of other snakes appears to have no effect on this snake.

About twelve eggs are laid in summer. These take some seventy days to hatch with young measuring 25 cm.

434 King Brown Snake *Pseudechis australis*, head (Winton, QLD).

COLLETT'S SNAKE
Pseudechis colletti Boulenger, 1902
Figs 436–439

Restricted to dry, non-swamp parts of inland Queensland, this brightly coloured and deadly snake attains 1.5 metres in length, although specimens of more than 2 metres are known. Juveniles are considerably brighter than adults, having more red or orange coloration present.

The scalation is smooth with 19 mid-body rows, 215–235 ventrals, divided anal, and 50–70 subcaudals, some single and some divided.

The Collett's Snake is very docile in nature and rarely bites. Its venom is neutralised by *Pseudechis papuanus* antivenom. If harassed, this snake may flatten out its entire body. It is a diurnal species. This snake is an opportunistic feeder, feeding on a variety of vertebrates including other snakes.

This species produces about twelve eggs to a clutch. Hatchlings measure about 37 cm. This species is particularly easy to keep in captivity and breeds on a regular basis.

435 King Brown Snake *Pseudechis australis* (Eyre Peninsula, SA).

436   Collett's Snake *Pseudechis colletti* (QLD).

437   Collett's Snake *Pseudechis colletti*, head (QLD).

438   Collett's Snake *Pseudechis colletti* (Qld).

439   Collett's Snake *Pseudechis colletti*, juvenile (Julia Creek, QLD).

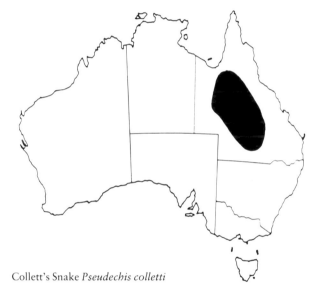

Collett's Snake *Pseudechis colletti*

BLUE-BELLIED OR SPOTTED BLACK SNAKE
Pseudechis guttatus De Vis, 1905
Figs 440–441, 541

This snake varies in colour from black, to grey or whitish, with variable markings. It attains 1.5 metres and is found in drier habitats of the north-eastern quarter of New South Wales and south-eastern Queensland. Although no deaths are recorded from the bite of this snake, it is fairly excitable and has potent venom, so it should be regarded as dangerous.

The scalation is smooth with 19 mid-body rows, 175–205 ventrals, divided anal and 45–65 single and divided subcaudals.

This diurnal snake is more pugnacious than other members of the Black Snake family, hissing loudly and flattening out its body when disturbed. It occurs in a range of drier habitats, although in very dry areas this snake is usually found near water.

The Blue-bellied Black Snake feeds on small mammals, frogs and lizards. It produces seven to thirteen eggs. Incubation takes eleven to twelve weeks with hatchlings measuring about 28 cm.

Blue-bellied Black Snake
Pseudechis guttatus

RED-BELLIED BLACK SNAKE
Pseudechis porphyriacus (Shaw, 1794)
Figs 442, 443

This distinctive dangerous snake is found in wetter parts of south-eastern and eastern Australia. It is found along all parts of the Murray–Darling river system. In Queensland this species has been decimated by the introduced Cane Toad *Bufo marinus*. It averages 1.5 metres although specimens (usually males) more than 2 metres long are known.

The scalation is smooth with 17 mid-body rows, 180–210 ventrals, divided anal and 40–65 single and divided subcaudals.

Although the Red-bellied Black Snake has killed people, its venom is not nearly as potent as is widely believed. Typical symptoms of this snake's bite are much local pain and swelling, nausea and general sickness. Fortunately this snake is very inoffensive and rarely bites. When agitated and cornered it may flatten its neck and raise its head, not unlike a Cobra *Naja naja*.

This diurnal snake is most commonly found in the vicinity of water, which it may frequently enter in search of its food. An opportunistic feeder, it feeds mainly on frogs, lizards and small mammals, although it has been known to climb trees, presumably in search of birds.

Red-bellied Black Snake
Pseudechis porphyriacus

440 Blue-Bellied Black Snake *Pseudechis guttatus* (Hunter Valley, NSW).

441 Blue-Bellied Black Snake *Pseudechis guttatus*, one-year-old juvenile (Moonie, QLD).

442 Red-Bellied Black Snake *Pseudechis porphyriacus* (Seven Hills, NSW).

Springtime aggregations of this snake occur, as does male combat. When fighting one another, Black Snakes are effectively oblivious to all that goes on around them. Rarely, if ever, do they harm one another in combat.

In summer from eight to thirty live young are born in membraneous sacs, from which they emerge minutes after birth. Newborns measure about 18 cm.

In captivity this species is easy to maintain, being a voracious feeder and resistant to ailments.

443   Red-Bellied Black Snake *Pseudechis porphyriacus*, head (Seven Hills, NSW).

444   Western Brown Snake *Pseudonaja nuchalis* (Tiboburra, NSW).

445   Western Brown Snake *Pseudonaja nuchalis*, head (Tiboburra, NSW).

447   Western Brown Snake *Pseudonaja nuchalis*, juvenile (Cobar, NSW).

446   Western Brown Snake *Pseudonaja nuchalis* (Windorah, QLD).

WESTERN BROWN SNAKE
Pseudonaja nuchalis Gunther, 1858
Figs **444–447**

This swift-moving and highly dangerous snake is found in most parts of Australia, except for the far south-east, east coast and most of the far south coast of South and Western Australia. The colour of this 1.6-metre snake is highly variable, ranging from whitish, reds, yellows, browns and almost black in colour, with or without head and body markings. In some areas this snake is commonly confused with the Inland Taipan, which it superficially resembles.

The scalation is smooth with 17–19 mid-body rows, 180–230 ventrals, divided anal and 50–70 divided subcaudals.

This diurnal snake is found in all types of habitat, and is common in farming country, where it feeds on introduced mice and rats as well as native vertebrates. Although aggressive, when disturbed this snake is not as aggressive as the Eastern Brown Snake.

Male combat in this species occurs, and about twenty eggs are produced. Young specimens almost always have head markings, which may or may not fade with age. The hatchlings measure 22 cm.

Western Brown Snake *Pseudonaja nuchalis*

448   Eastern Brown Snake *Pseudonaja textilis* (Cobar, NSW).

EASTERN BROWN SNAKE
Pseudonaja textilis (Dumeril, Bibron and Dumeril, 1854)
Figs 21, **448–456,** *507–530*

This swift-moving snake is a close relative of the Western Brown Snake *Pseudonaja nuchalis*. The two species are easily confused and commonly occur in the same areas. The Eastern Brown Snake has a flesh-coloured buccal cavity, as opposed to a blackish colour found in the Western Brown Snake. The 1.6-metre, highly dangerous Eastern Brown Snake is found throughout the eastern half of Australia, its distribution becoming patchier as one moves westwards. To date only one specimen has been caught in Western Australia.

The colour of this snake ranges from near white through various colours to jet black. Some black specimens result from a specific allele (type of gene), and black and non-black specimens may result from a single clutch of eggs. Juveniles from the coast of New South Wales are strongly banded (illustrated), while those from elsewhere typically have markings on the head only (also illustrated). In both cases these markings usually fade with age. In borderline areas both banded and unbanded snakes may emerge from the same clutch of eggs.

The scalation is smooth with 17 mid-body rows, 185–235 ventrals, divided anal and 45–75 divided subcaudals.

This species has extremely toxic venom, but fortunately its biting apparatus is not as well developed as in most other deadly snakes. It injects relatively little venom in most bites and its fangs are relatively short, although they can still easily penetrate the skin. Brown Snakes are fast-moving and potentially highly aggressive. When aroused a Brown Snake will hold its neck high, slightly flattened in an S-shape, and strike repeatedly at its aggressor. This snake will occasionally chase an aggressor away, striking at it at every opportunity.

The diurnal Eastern Snake is most common in dry grassy country with scattered ground cover, but occurs in all types of habitat. When resting this snake utilises any

449   Eastern Brown Snake *Pseudonaja textilis*, ventral surface (Cobar, NSW).

450   Eastern Brown Snake *Pseudonaja textilis*, head (Windsor, NSW).

451   Eastern Brown Snake *Pseudonaja textilis*, melanistic specimen (Green Valley, NSW).

452   Eastern Brown Snake *Pseudonaja textilis*, melanistic specimen, ventral surface (Green Valley, NSW).

453   Eastern Brown Snake *Pseudonaja textilis*, juvenile (St Clair, NSW).

454   Eastern Brown Snake *Pseudonaja textilis*, juvenile, head (St Clair, NSW).

455   Eastern Brown Snake *Pseudonaja textilis*, two-year-old juvenile (St Clair, NSW).

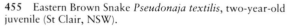

Eastern Brown Snake *Pseudonaja textilis*

170

available cover, but is particularly fond of man-made cover such as sheets of metal.

Diet is varied, but introduced pest rodents are a preferred item.

This species has large winter aggregations, with one consisting of thirty individuals being found near Sydney. These aggregations are maintained in spring for mating purposes. Males, which are often the larger sex, also engage in combat. Ten to thirty eggs are produced in summer, which hatch about eighty days later. Hatchlings measure about 27 cm.

HALF-GIRDLED SNAKE
Simoselaps semifasciatus (Gunther, 1863)
Figs 457, 458

Found throughout the western half of Australia and north-eastern Queensland, there are several species covered under this name. This snake is variable in colour. Two variants are pictured here. Average length is about 35 cm.

The scalation is smooth with 15–17 mid-body rows, 140–190 ventrals, divided anal and 14–30 divided subcaudals.

This inoffensive snake is usually found when crossing roads at night. If caught during the day, it is usually found under ground litter such as logs, well-embedded rocks, etc. The Half-girdled Snake occurs in all types of habitat but is most common in drier areas.

It appears that the diet of this snake consists exclusively of reptile eggs. This species is itself an egg-layer, although its biology is little known.

Half-girdled Snake *Simoselaps semifasciatus*

CURL OR MYALL SNAKE
Suta suta (Peters, 1863)
Figs 459, 460

Found in most drier parts of the eastern two-thirds of Australia, this 50-cm snake is generally a brownish colour, with reddish or other tinges.

The scalation is smooth with 19–21 mid-body rows, 150–170 ventrals, single anal, and 20–35 single subcaudals.

The nocturnal Curl Snake gets its name from its defensive posture. It will flatten out its body into a rigid spring-like curve, from which it will strike if necessary.

During the day this terrestrial snake is found under

456   Eastern Brown Snake *Pseudonaja textilis*, juvenile (Bathurst, NSW).

457   Half-girdled Snake *Simoselaps semifasciatus* (Lake Argyle, WA).

458   Half-girdled Snake *Simoselaps semifasciatus* (Shay Gap, WA).

ground litter, while at night it is commonly found crossing roads. It occurs in all habitats except for the wettest forests.

The Curl Snake is an opportunistic feeder, feeding principally on small lizards and small mammals.

Two to seven live young are produced in summer.

459   Curl Snake *Suta suta* (Bourke, NSW).

Curl Snake *Suta suta*

460   Curl Snake *Suta suta* (Boggabri, NSW).

461   Rough-scaled Snake *Tropidechis carinatus* (Barrington Tops, NSW).

ROUGH-SCALED SNAKE
Tropidechis carinatus (Krefft, 1863)
Figs 461–463

This relatively aggressive and dangerous snake is commonly confused with the Keelback *Amphiesma mairii*, which it resembles superficially. The Rough-scaled Snake is found in rainforest areas of northern New South Wales from Barrington Tops to Cape York in Queensland, and attains about 75 cm in length. I caught a specimen of 1.2 metres at Mount Nebo, Queensland.

The scalation is strongly keeled, hence the name 'Rough-scaled Snake', with 23 mid-body rows, 160–185 ventrals, single anal and 50–60 single subcaudals.

This nocturnal rainforest dweller is occasionally seen active during the day in colder months. Most specimens are caught when crossing roads at night. This snake is usually very abundant where it occurs. It feeds on various vertebrates, particularly frogs and lizards, and this species has suffered greatly from the introduced Cane Toad *Bufo marinus*.

This snake is highly aggressive, and captive specimens are no less aggressive, even after several years in captivity.

About five to eight live young are produced.

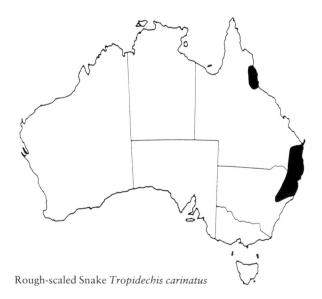

Rough-scaled Snake *Tropidechis carinatus*

462 Rough-scaled Snake *Tropidechis carinatus* (Jellaten, QLD).

463 Rough-scaled Snake *Tropidechis carinatus*, juvenile (Mount Glorious, QLD).

CARPENTARIA WHIP SNAKE
Unechis boschmai (Brongersma and Knaap-van Meeuwen, 1964)
Figs 464, 465

Attaining 40–50 cm, the Carpentaria Whip Snake is found mainly in the hilly country to the west of the Great Divide in Queensland, although it does occur in drier habitats east of the ranges. This snake may be tan, brown, or reddish in colour.

The scalation is smooth with 15 mid-body rows, 145–190 ventrals, single anal and 20–35 single subcaudals.

This secretive nocturnal snake is usually found hiding under ground litter during the day. It is essentially a small-lizard-feeder and bears live young. Although freshly caught specimens will attempt to bite, captive specimens are very inoffensive.

464 Carpentaria Whip Snake *Unechis boschmai* (Dalby, QLD).

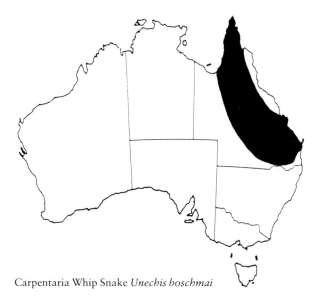

Carpentaria Whip Snake *Unechis boschmai*

HOODED SNAKE
Unechis monachus (Storr, 1964)
Fig. 466

This 40-cm brownish or greyish snake is common in drier parts of southern Australia.

The scalation is smooth with 15 mid-body rows, 150–175 ventrals, single anal and 20–35 single subcaudals.

This species is found in forests and all other types of drier habitat. The Hooded Snake is a secretive nocturnal

465 Carpentaria Whip Snake *Unechis boschmai*, head (Dalby, QLD).

species, usually caught during the day hiding under ground litter. It is principally a lizard-feeder.

A live-bearer, little is known of the breeding biology of this species.

173

Hooded Snake *Unechis monachus*

466   Hooded Snake *Unechis monachus* (Cobar, NSW).

467   *Unechis spectabilis* (Crystal Brook, SA).

Unechis spectabilis (Krefft, 1869)
Fig. 467

Unechis spectabilis attains 30 cm and is found in southern Queensland, most of New South Wales, north-western Victoria, southern South Australia and adjacent parts of Western Australia. Specimens from Queensland and most of New South Wales are sometimes classed as a separate species, namely *Unechis dwyeri*.

The scalation is smooth, with 15 mid-body rows, 135–70 ventrals, single anal and 20–40 single subcaudals.

Unechis spectabilis is a secretive species which is usually found during cooler weather hiding under ground cover. Nocturnal and only found active in warmer weather, this species is believed to feed on small lizards.

The breeding biology of *Unechis spectabilis* is not known.

Unechis spectabilis

BANDY-BANDY
Vermicella annulata (Gray, 1841)
Figs 468, 469

Found in about half of mainland Australia, excluding the west and far south-east, the Bandy-Bandy averages 60 cm although specimens of 80 cm are known. Similar, closely related species are found in other parts of Australia.

The scalation is smooth with 15 mid-body rows, 180–230 ventrals, divided anal and 14–28 divided subcaudals.

The Bandy-Bandy is found in all types of habitat, from rainforests to deserts. It is usually found during the day under cover or on mild nights moving around above the ground surface. The preferred night surface activity temperature appears to be lower than that of most other snakes, and is more in line with Blind Snakes (*Typhlopidae*), its principal food. This snake consumes very little food over very long periods because of its unusually slow metabolic rate.

When caught this snake is usually nervous, often knotting itself around one's hand in a manner not unlike that of the Blind Snakes (*Typhlopidae*). If suddenly alarmed this snake may flatten its body and elevate parts in loops in a bluff display. This display position can be maintained for some time.

This snake produces four or five eggs in the warmer months. Hatchlings measure 17 cm.

468 Bandy-Bandy *Vermicella annulata* (West Head, NSW).

469 Bandy-Bandy *Vermicella annulata*, head (West Head, NSW).

Bandy-Bandy *Vermicella annulata*

470 Spine-bellied Sea Snake *Lapemis hardwickii* (Hervey Bay, QLD).

Family Hydrophiidae (Sea Snakes)

The Sea Snakes are specifically adapted to a marine existence. They have vertically flattened tails, valvular nostrils and other distinctive characteristics. These snakes are believed to have originated from the same stock as modern Elapids as they are also front-fanged and in general highly venomous. Little research has been done to date on the toxicity of Sea Snake venoms, although most species are believed to be potentially dangerous. No Sea Snakes are aggressive, and considering that large numbers of fishermen come into contact with sea snakes on a daily basis, bites are rare.

Sea Snakes are able to rapidly dive deeply and surface again without getting that human affliction called bends caused by air bubbles forming in the bloodstream.

With the exception of one species, Yellow-bellied Sea Snake *Pelamis platurus*, which is found throughout the Indian and Pacific oceans, Sea Snakes are restricted to South-east Asia and Australia (Persian Gulf to Japan). About fifty species have currently been described, of which more than thirty are recognised as coming from Australian waters. Most Australian varieties are only found in northern waters.

All except two species are live-bearers, and those two species have been placed by some researchers in a separate family (Laticaudidae), as done earlier in this book.

SPINE-BELLIED SEA SNAKE
Lapemis hardwickii Gray, 1835
Figs 470, 471

Found throughout the northern waters of Australia in a line north of Brisbane, specimens have been known to stray as far south as Sydney. This dangerous snake is one of the largest species of Sea Snake, and grows to 1.3 metres in length.

The scalation is juxtaposed with 23–45 mid-body rows and 110–240 very small ventrals.

175

471   Spine-bellied Sea Snake *Lapemis hardwickii*, head. Note the dorsal positioning of the nostrils and the 'valves' present. (Hervey Bay, QLD).

472   Yellow-bellied Sea Snake *Pelamis platurus* (Manly Beach, NSW).

This common Sea Snake is usually seen on the surface of both coastal and deep waters. It is commonly caught by trawlers in their nets, throughout northern Australia. Like all Sea Snakes, this species rarely attempts to bite.

This snake is a fish-feeder. Captive specimens have been known to live for up to ten years and appear to be resistant to most types of ailment.

A live-bearer, this species' biology is little known.

monly found in waters along the entire west, north and east coasts. This dangerous species grows to about 70 cm.

The scalation is imbricate with 47–69 mid-body rows and 264–406 ventrals.

This snake has killed a number of people, although to date no Australians are known to have died from its bite. The venom is strongly neurotoxic (nerve-killing).

No Sea Snake can cross land well, but this species can do so better than most and so if found on a beach it should be treated with extra caution.

Yellow-bellied Sea Snakes feed on small fish and young eels. This species is known to occur in large aggregations in some areas, numbering hundreds or even thousands of individuals. The purpose of these aggregations is not known.

This species produces from one to six live young, although its breeding biology is essentially unknown.

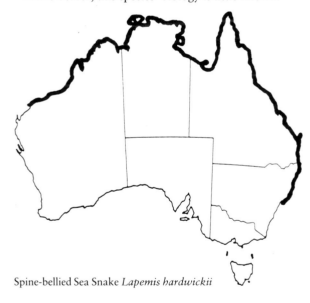

Spine-bellied Sea Snake *Lapemis hardwickii*

YELLOW-BELLIED SEA SNAKE
Pelamis platurus (Linnaeus, 1766)
Fig. 472

Commonly called the Pelagic Sea Snake, this is the only truly ocean-going species of Sea Snake. It is found throughout the Indian and Pacific oceans and is com-

Yellow-bellied Sea Snake *Pelamis platurus*

PART 3   CAPTIVITY AND CONSERVATION

I ADVISE THOSE who intend to keep, or who already keep, reptiles and frogs to read all the following sections of this book, in particular the section on conservation. Throughout most of the text following I deal in generalisations that apply to most species but not to all species referred to. This section should be read in conjunction with the descriptions of habits of any particular species given in Part 2. The reader must also seek further information elsewhere if needed. Above all, one should maintain a conservationist and humanitarian attitude to all reptiles and frogs, on an individual and species level.

OBTAINING REPTILES AND FROGS

TO OBTAIN A given reptile or frog one can either obtain it from a captive source such as a dealer or keeper, or catch it oneself. Currently in Australia there are few reptile keepers, fewer breeders and virtually no dealers in any form. This means that in order to obtain reptiles and frogs most people will have to catch their own in the wild state.

To obtain a given species the easiest way is to find out about its habits, distribution, etc, and then act accordingly. Anyone within Australia can (legally) obtain any reptile species found within Australia if they apply themselves to the task, and have justification in doing so.

When collecting reptiles and frogs from the wild it is important to keep detailed records of what is found. This includes the following information:

1 Date.
2 Time of day.
3 Map location.
4 Numbers and species caught.
5 Weather conditions.
6 Where specimens were found.
7 Collector's name(s).
8 Other relevant information.

Very importantly, when collecting, the following points should be remembered:

1 Minimise environmental impact or destruction. In other words replace rocks, logs, etc. as found where possible. Don't move anything that doesn't need moving.
2 Never collect more specimens than required.
3 Have adequate means to transport captured specimens.

Finding the Reptiles and Frogs

Most reptiles and frogs appear to be more prevalent or visible at certain times of the year. Finding out these times of the year for the desired species makes the task of collecting them considerably easier.

Frogs

Generally frogs are found near water, and may be found during the day resting in vegetation around creeks, swamps, dams, etc, or under ground cover nearby. Most specimens are caught at night when they are active, particularly on wet nights. Because males call, large congregations of breeding frogs can be found at certain times of the year by following their calls. Usually where frogs are calling individuals can be located without too much difficulty.

However, some frogs appear to be ventriloquial and are difficult to locate even when very close. 'Triangulation' is a method of locating these frogs. By two or more people from opposite directions moving towards the point where they think the frog is calling, their paths or light beams should intersect the point where the frog is. Collectors need to practise triangulation for it to work well, as well as having considerable patience when frogs are reluctant to call.

Calling frogs that are hard to locate can be found with the use of a cassette player. By recording a frog(s) calling and playing it back in the same place immediately after taking the recording, specimens of the same species may be attracted to the cassette recorder in search of a mate.

Frogs will be found crossing roads on wet nights in suitable habitats and feeding on insects at most times of the year.

Tadpoles of frogs are usually found in unpolluted bodies of water without many fish present. This includes most dams, small creeks, swamps, etc. Usually tadpoles are found in very large numbers, so collecting them rarely poses difficulty.

Reptiles

Reptiles are found in virtually all habitats in Australia. Usually, but not always, virgin habitats have the most reptiles, both in terms of species and of individuals. Habitat types which virtually always have large numbers of reptiles include rocky habitats, arid regions covered

473   Capture of a Death Adder *Acanthophis antarcticus* crossing a road during unstable weather in daytime, at Mount Glorious, QLD. Driving along suitable roads, particularly at night, is one of the most effective ways to locate herptiles.

with spinifex *Triodia* spp., areas adjacent to watercourses, and other habitats with abundant ground cover.

In most parts of Australia reptiles may be found under ground cover during cooler weather conditions. When lifting ground cover some species (particularly geckoes) have a tendency to hold on to the underside of the cover. This should be watched for so that: (1) you don't miss the specimen, and (2) you don't squash the reptile when replacing the cover.

Other burrowing reptiles and frogs may shelter under leaves or soil under ground cover and in some circumstances it is worthwhile to rake soil, leaves, grass, etc., under cover to find these species.

Rock crevices are a favourite hiding-place for many species, including Tree Snakes, Pythons and spiny *Egernia* Skinks, although specimens found in crevices are often almost impossible to capture.

Arboreal species may also be found under tree bark (e.g. most *Oedura* geckoes), or hollow tree limbs (e.g. Spotted Tree Monitor *Varanus timorensis*).

Sheet iron and other man-made rubbish is preferred cover for many reptiles. Places such as country rubbish tips and cemeteries are often excellent for finding reptiles.

Active diurnal reptiles are usually seen during the day when moving through their habitat. When walking, by listening carefully one can often hear reptiles fleeing through ground litter as one approaches them. In sandy areas one can follow reptile tracks to eventually locate an active or resting reptile.

In arid areas many lizards dig and occupy burrows. These usually have their entrance at the edge of a rock or spinifex (*Triodia* spp.) bush. There is often a second entrance or especially constructed 'emergency exit' just below the ground surface from which a pursued lizard may flee. One should be aware of this when chasing these lizards. The burrows of large scorpions can be confused with those of lizards. Usually scorpion burrows have U-shaped openings as opposed to the n-shaped openings of lizard burrows.

Night driving in warm weather is without doubt the most effective way to find many nocturnal reptiles. Reptiles will be found crossing roads running through suitable habitat. At night the open roads become effectively just another strip of bush. By covering a wide strip of bush at high speed in a car, one is able to have a much greater chance of finding moving reptiles than would otherwise be possible by walking through the same habitat with a spotlight or torch.

By driving along roads during the day, through certain areas at particular times of year, one may find reptiles (mainly lizards) crossing, or be able to spot agamids perched on fenceposts along the sides of the road.

In many areas reptiles have a habit of falling down mine shafts and other man-made holes in the ground, and then remaining trapped in these holes until they starve to death. Checking these holes can often reveal numbers of trapped reptiles.

Tortoises are found in most rivers and larger watercourses throughout Australia. In clear waters the best way to locate specimens is by diving for them with a mask, flippers and a snorkel. Where waters are too murky for this, nets and fish traps are a highly effective way to capture tortoises. It should be mentioned that tortoises will drown in nets, so these should be checked frequently to avoid this risk. Traps should be set in such a way that when the tortoise is trapped it can still swim to the water surface, again to prevent drowning.

Crocodiles may also be found using the same techniques as for tortoises, although most specimens are seen at night with the aid of spotlights from boats.

Marine turtles may be located by going to their nesting beaches, which are usually well known, during the breeding season. The adults are hard to miss because of their size, slow-moving nature and the tracks they leave. Hatchlings are attracted to human lights placed on the beach on the night that they hatch, and if a beach is well covered with people with torches, hatchlings won't be missed. Fishermen often capture sea turtles in nets, and divers also see them in tropical seas. Sea snakes are also found in large numbers by trawlers, and regularly seen by divers in warmer waters.

The use of trained dogs to find reptiles has been little used by reptile collectors. For a number of years I had a pet dog, a Dachshund/Doberman cross, which was well trained at finding reptiles by following their scent trails. This dog's efficiency in locating large lizards and snakes equalled that of about ten people. When the dog found a reptile, it simply barked loudly until I came and caught the specimen. On one occasion the dog found thirty-three Cunningham's Skinks *Egernia cunninghami* near Oberon, New South Wales, in half an hour.

Capturing Reptiles and Frogs

Most reptiles and frogs are usually ambushed as they try to make an escape, and are simply 'grabbed'. A number of tools used to capture reptiles and frogs have already been listed, including **cars** for road hunting, **spotlights** for frogs and some nocturnal reptiles, **cassette player** for attracting frogs, **trained dog** for terrestrial reptiles, **fishing gear** including **nets** for aquatic reptiles and tadpoles, and **diving equipment** for other aquatic reptiles.

When attempting to capture a reptile, most collectors will improvise and use available aids as necessary, including some of the following:

Crowbars and **jacks** are useful in lifting large, otherwise immovable rocks when reptiles are known to be hiding under them.

Prodding wires can be used to poke reptiles out of the rock crevices by prodding them in sensitive areas such as the neck and behind the limbs. Often collectors will use the nearest stick to poke a reptile from a crevice.

Fire is useful in getting reptiles to flee from cover, either through burning the cover or by **smoking** the reptile out of its shelter.

Shovels and **rakes** are useful for digging out burrows and sifting through leaf litter when a reptile has taken cover. **Axes** and **saws** are useful for extracting reptiles from logs and tree branches, although they are highly destructive.

Rubber bands and **sling shots** can be used to stun lizards that won't allow one to approach too close.

Guns can be used to obtain reptiles that are not necessarily needed alive (for example, for a museum collection), although the use of guns for capturing or killing reptiles should generally be discouraged.

A **noose** is useful for obtaining otherwise inaccessible reptiles such as Goannas in trees. A noose is essentially a loop of rope at the end of a pole that will tighten when pulled. When the loop is slipped over the lizard's head and yanked, it tightens, allowing the lizard to be pulled from the tree. When in the middle of bushland with few tools a noose can be made from a long stick with a piece of string at one end with a slip knot tied.

Traps, such as modified 'possum traps' which rely on bait to capture a reptile, may be used for larger monitors, but usually with limited success. Smaller bait traps for skinks have been used with limited success by some collectors.

Pit traps, which are specially-dug holes into which reptiles fall, are usually of minor success. These traps are in the path of reptiles. **Drift fences** or **barriers** may be used to direct reptiles towards the pit traps, which may in turn be baited with food for the lizard to be caught. Another disadvantage of these traps is that they need to be: (1) specifically constructed in the first place, and (2) checked frequently to be at all successful.

Snake sticks (including **jiggers**) come in a variety of forms. The traditional forked stick used to pin snakes by the head is generally not used because: (1) it can injure the snake, and (2) it is unlikely that a stick on hand will be the same size as the snake.

Modern snake sticks (jiggers) consist of a fork at the end, with a stiff rubber across the fork. Snakes can be pinned with this apparatus and then picked up relatively safely. Variations of this form do occur. A stick with a hook at the end is rarely useful for caturing wild snakes, although it is excellent for handling captive snakes which are more docile and less likely to flee when handled.

A number of other tools for capturing reptiles and frogs are sometimes used, and the variety is only limited by the innovativeness of reptile and frog collectors.

Handling Reptiles and Frogs

When handling reptiles and frogs the two main considerations are the welfare of the animal and one's personal safety.

Handling of any newly-caught herptile (reptile or frog) should be kept to a minimum, as excessive handling may distress or physically harm the animal. Frogs have a soft permeable skin, and may actually be burned by the warmth of human hands. Humans may occasionally suffer skin reactions from the secretions of frogs. All herptiles should be handled gently but firmly, and with support given to the body, and they should never be allowed to escape through loose grip on the part of the handler. The handler's grip should be such that the herptile is unable to change its positioning in any way. Otherwise the handler is at risk, and that is not an acceptable situation.

Most lizards have autotomy, the ability to shed their tails, and this should be prevented at all costs, for a number of reasons. Therefore it is advisable never to handle any lizard by the tail. Also if a lizard is trying to 'throw' its tail around when being handled, this should be brought under control.

In most cases any reptile that is likely to give a painful or poisonous bite should be firmly held behind the head so that it cannot turn around and bite the handler. Snakes are first pinned by the head on the ground with a jigger stick before being picked up by the handler by the back of the head. Gloves should not be worn by reptile handlers as the sensitivity lost through the gloves actually increases the risks of a bite, through possible bad handling. (In a few cases gloves may be useful when handling small venomous snakes, where it may be difficult to avoid a bite by any other form of handling.)

Some poisonous snakes may be 'tail-handled' because of the difficulty of grabbing them behind the head. In this case the snake is grabbed by the tail region (always just above the vent) and its body is twisted constantly so that the snake is always out of balance and therefore unable to rise up and bite the handler. Usually most venomous snakes are handled in this manner and then thrown into an open bag which is then rapidly closed and sealed.

Hooks, which are often used to lift and carry venomous and other snakes in captivity, are useful in that they minimise contact between handler and snake, thereby minimising the risk of a bite. Captive snakes often become used to being 'hook-handled' and become very co-operative when being handled in this manner.

Lizards with sharp claws, such as monitors, should be held away from one's own body so that they can't inflict severe scratches. Larger lizards are capable of drawing blood with their claws. Long muscular whip-like tails of larger monitors and some agamids can also cause problems and should therefore always be held away from the face and other sensitive areas.

Testudines generally pose no handling problems although the jaws of larger specimens should be watched closely.

Medium-sized to large crocodiles should be handled by more than one person, and often need to be 'roped' or tied up before actually being handled. They have large, powerful and potentially destructive tails, not to mention their jaws. Captive specimens are often more dangerous than wild specimens.

Transporting Reptiles and Frogs

When transporting reptiles it is most desirable to avoid putting reptiles of disparate sizes together, as the larger specimens may injure or crush the smaller ones. Also overcrowding of reptiles should be avoided, and when possible reptiles should be transported in individual bags or containers.

Most reptiles are best transported in cloth bags. Bank bags, pillow cases and flour bags all make excellent snake bags, which are tied with string at the top when containing reptiles. The advantages of bags for carrying reptiles are many and include the following:

1 Reptiles are unlikely to injure themselves (particularly their snouts) when trying to escape. This is much more likely when reptiles are being transported in cages or other containers.

2 The bag minimises stress for the reptile. It is shielded from seeing movements outside the bag and is therefore less likely to become alarmed when being transported.

3 The bags themselves fold into a small flexible package when not being used to hold reptiles, and are frequently useful for other purposes.

When reptiles are being carted in separate bags within a single container, the bags will act as a weight buffer between specimens in each bag. Where larger heavier specimens are being transported in bags, it is best to separate each individual bag when transporting. This can be done by various means, including hanging each bag from a hook, or by placing each bag within a separate box.

Reptiles including snakes are excellent at making small holes into bigger ones and escaping, so bags in use should be regularly checked, and venomous reptiles should be 'double bagged'.

Smaller specimens can be carried in jars and other similar containers, usually with some leaf litter or tissue paper inside to act as cover for the reptile and prevent them from being shaken and bumped when transported.

Frogs may be transported in similar ways to reptiles except that it is best to keep them moist at all times. Plastic bags are highly desirable for frogs as they do not need constant moistening. As most frogs have toxic secretions, it is exceptionally important to avoid crowding, and mixing of species, when transporting. If travelling for any substantial period, it may be necessary to wash the frogs and their bags or containers to remove toxic secretion build-up.

Eggs of all reptiles and frogs should be transported with minimum movement, and it should be remembered that changing the orientation of an egg (e.g. turning it upside down) or shaking it can kill the embryo.

The greatest enemy of reptiles and frogs being transported is heat and associated dehydration. Therefore it is always important to keep transported specimens as cool as possible. Here common sense plays a major part. Never leave bags or containers in direct sunlight and avoid placing them on warm ground. It goes without saying that reptiles should never be left in stationary vehicles during the day; if need be, they should be placed underneath the vehicle when stationary.

Moist bags will remain cooler than dry bags in warm weather, and help prevent dehydration of specimens. Plastic bags and containers are more susceptible to overheating than cloth bags.

If there is any doubt as to which specimen is in a given container, labelling should be used.

KEEPING REPTILES AND FROGS

The Cage

Most reptiles and frogs are best kept in glass-fronted cages, such as fish tanks, converted packing-cases, shop display cabinets, or especially constructed cages. The minimum size of the cage should reflect the needs of the specimen(s) to be housed in it. For example, arboreal species usually require a tall cage while for terrestrial species cage height is usually unimportant. Although there are no strict rules for cage size, I will give examples of what I view as minimum cage size for reptiles and frogs (see Table 1).

In some parts of Australia it may be possible to house larger specimens in outdoor pits. These usually are walled enclosures with a large landscaped area that allow the reptiles to enjoy near-natural surroundings.

One should not house incompatible herptiles in the same cage. For example, one should not house lizard-eating snakes and lizards together. As a general rule individual species should be kept separately; however, similar species often may be kept together in larger cages. It is not uncommon to keep a number of types of larger lizard or tortoises in outdoor cages or pits without any problems.

When housing reptiles and frogs it is important to spare no expense in providing the best and most suitable accommodation for the specimens concerned as failure to do so will invariably result in premature death of specimens.

When constructing a cage for herptiles there are a number of essential points:

1 The cage must be escape-proof. Most herptiles are experts at escaping, so extra care is needed.

2 The cage must be located in a position where it is relatively secure from burglaries and theft of specimens. Outdoor reptiles must be secure from cats, birds, etc.

3 The keeper must have total access to all specimens at all times in case of diseases, etc.

4 The enclosure must have an environment that will enable the specimen to maintain good health. This usually includes:
  a The cage should be of adequate size for the specimen to be placed in the cage. It is better to build a larger cage if in doubt.
  b Fresh clean water must always be available, usually in a non-spillable, cleanable dish or bowl. The reptiles or frogs should always be able to immerse themselves completely in water if they so choose.
  c Adequate cover for the specimen.
  d Required interference by the keeper for maintainence should be kept to a minimum, as herptiles should not be disturbed unnecessarily.
  e The enclosure should afford the herptile adequate protection from climatic extremes, and provide opti-

Type of herptile	Minimum sizes of cage Length × width × height (metres)
Frog	0.5 × 0.3 × 0.3
Crocodile	5 × 5 × N/A
Sea turtle	10 × 8 × 1.5 (water only)
Freshwater tortoise (under 30 cm)	1 × 1 × 0.5 (pond only) 1.5 × 1 × 0.5 (total cage)
Geckoes, pygopids and small skinks	1 × 0.5 × 0.4
Small dragons and small monitors (av. 30 cm)	1 × 1 × 0.5
Large skinks, large dragons and medium monitors (av. 60 cm)	2 × 1.5 × 1
Large monitors (over 1.1 metre)	4 × 3 × 2
Snakes under 0.5 metre	1 × 0.5 × 0.5
Snakes over 0.5 metre but under 2 metres	2 × 0.5 × 0.5
Snakes over 2 metres	3 × 0.5 × 0.5

Notes

1 This table consists of generalisations only. The operative word is *minimum*. Most herptiles will do best in larger cages.
2 The above cage sizes are quoted for one or two adult specimens only, and if one intends putting more than two individuals in a cage, then the cage should probably be larger.

Table 1: Generalised examples of minimum cage sizes for reptiles and frogs

474 Author's outdoor pits immediately after construction. Note the wire above the pit to give total security from predators and theft, and to make the cages completely escape-proof.

475 View inside author's outdoor pit two years after completion.

476 View of author's outdoor pit two years after construction.

mum temperatures for the herptile at most times. Heating or cooling systems may be required.

The environment provided should be conducive to the herptile settling into captivity. Although some snakes, particularly Pythons, can be kept successfully in very spartan cages, when in doubt it is best to landscape a cage in a manner as similar to the natural habitat as is possible.

A number of aids in making a cage suitable for herptiles may be used. These include filters and pumps to provide clean water. Heating cables, light bulbs, 'hot rocks' and 'hot boxes' can all be used to provide warmth to specimens if needed. Air-conditioning units may be used to heat or cool reptile cages. Thermostats are useful in controlling cage temperatures when temperature control is needed but one cannot be present at all times.

Various lights may be used to provide daylight for indoor reptiles. Trulite is the form of artificial light most like natural sunlight and is consequently widely used by reptile keepers. Agamids in particular need sunlight or similar to survive in captivity.

Water spray systems may be used to increase humidity in cages if necessary (usually for frogs).

Frogs

Frogs are best kept in moist humid environments, with plenty of litter, etc. A water spray system could be used to increase humidity. Tree frogs often thrive in plant terrariums sold in shops. Ground-dwelling and burrowing species should have soil at least 10 cm deep in the cage. This should be moist but not waterlogged.

Fresh water is essential to frogs and many species seem to prefer a cage with circulating fresh water. A fish tank filter or pump is an excellent and inexpensive way to provide circulating water for frogs.

Tadpoles can be raised in any small water container with well-cured water. They usually thrive on lettuce boiled for 10–20 minutes, although a few species are carnivorous. When metamorphosis approaches it is necessary to place a projecting surface into the water, such as a large rock, so that the froglets can emerge; otherwise they will drown. The water container must also be made escape-proof for the froglets.

Crocodiles

Crocodiles are usually kept in outdoor cages in warmer areas. These are usually very large (more than 100 square metres), including large ponds or lakes, etc, which must be regularly cleaned, and are usually only affordable by zoos, crocodile farms or very wealthy individuals.

Testudines

Marine turtles must be kept in very large tanks with fresh circulating sea water. Again only zoos, aquariums and similar institutions can usually afford to build the necessary facilities for sea turtles.

Freshwater tortoises are among the hardiest of reptiles in captivity. They thrive in outdoor pits with large ponds. Larger ponds are easier to maintain than smaller ponds as it takes longer for the tortoises to foul them up. Care should be taken in constructing the pond to make sure that the surfaces are not abrasive to the shells of the tortoises. Shallow ponds often have problems of rapid algae build-up, and this should also be realised when designing ponds. Tortoises do best in clear water.

It is important to provide suitable nesting areas in tortoise cages.

In colder areas, and for young specimens, tortoises are best kept indoors in aquaria. They should have a platform or land area on which to bask if necessary.

Tortoises will attempt to eat any fish kept in the same tank.

Lizards

Smaller lizards are best kept indoors in cages. Outdoor pits are usually the easiest way to keep larger lizards. Lizards do best in cages with pently of cover in which to hide, and plenty of space in which to forage.

Diurnal lizards kept indoors must be provided with natural sunlight in order to survive for any substantial period of time. If it is impossible or impractical to provide sunlight, then Trulite or similar are the only suitable light that can be used for these reptiles. Trulite is difficult to obtain in Australia, although it is readily available in the United States and Europe.

Nocturnal lizards including geckoes do not need to have their cages exposed to any sunlight, although they must be exposed to near normal photo periods.

Species of lizard from arid areas will die if kept in a humid cage. Conversely species from wet habitats may have problems in an excessively dry or sandy cage.

Snakes

Snakes can generally survive better in smaller cages than most other reptiles. Most snakes do best in bone-dry cages, although they should still always have a water-dish provided.

Access to cages containing venomous snakes should be particularly good, so as to reduce the risk of bites when removing them from the cages, which will have to be done from time to time. The safest access to snakes' cages is usually from above, so this should be considered when constructing these cages.

Because of the difficulty of handling some snakes, simplicity is a key word for snake cages, as one does not want

477 Cages for breeding snakes, positioned in a 'battery' to minimise heating costs.

478 Excellent outdoor pit at Queensland Reptile Park. Note the smooth interior wall.

to destroy some elaborate set-up to remove a snake, only the spend hours putting it back together again.

Burrowing snakes should be provided with moist sand or soil in which to burrow.

FEEDING REPTILES AND FROGS

FEEDING IS IMPORTANT in maintaining health of herptile specimens. Many species feed only on live food and it is often necessary to breed or regularly capture food for them.

Usual food for given herptiles in captivity are given in Table 2.

Herptiles have slow metabolic rates, and in general should be fed not daily but periodically, depending on the size of the animal — larger specimens reguire feeding less frequently. Regularity of feeding also depends on the size of meals eaten at each feeding, and can be worked out on the basis of what is needed to maintain condition. It is inadvisable to feed excessively large amounts at a single feeding session as the risk of regurgitation (and associated problems) becomes important.

One should in general feed herptiles nearly as much as they will take. However, it will take some experience to work out how much this is, and therefore it is advisable to feed new herptiles all that they will take (but still not too much per feeding). The theory is that 'slightly hungry' herptiles will always be more consistent feeders than ones which go through gorging then starving phases. (Almost all herptiles will go through some periods of not feeding anyway.)

Uneaten food or prey should *always* be removed from the cage for several reasons, including the following:

1 Herptiles with constant exposure to food have a tendency to loose their appetites.

2 Uneaten food will decay and create further problems.

3 Uneaten prey (including mice and rats) might actually attack the herptile and kill it when it is resting. (Many inexperienced keepers loose Pythons to mice when the mice are left in the cage overnight.)

If a herptile is presented with food during an 'active' period, and it does not eat the food within three hours, one can fairly safely deduce that the food will not be eaten.

479 Captive New England Cunningham's Skinks *Egernia cunninghami* feeding on canned pet food in author's pit.

Type of herptile	Foods eaten
Most frogs	Live insects only, occasionally mice.
Crocodiles	Fish, most types of meat including chickens.
Sea turtles	Fish, occasionally seaweed.
Freshwater tortoises	Raw meat, tinned pet food, small fish, snails, etc.
Most geckoes	Insects only.
Most pygopodids	Other lizards, occasionally fruit, such as banana.
Most dragons	Arthropods, live worms, occasionally plant material.
Small skinks	Live insects, occasionally meat and/or fruit.
Large skinks	Meat, including tinned pet food, fruit.
Monitors	Meat, other vertebrates.
Blind snakes	Unknown; presumably ants and/or termites.
Pythons	Mice, rats, chickens, etc.
Colubrids	Frogs, sometimes mice and small rats.
Most small elapids (60 cm or less)	Small lizards only.
Most medium-sized elapids (50 cm to 1 metre)	Small lizards, and/or mice, small rats, etc.
Most large elapids (1 metre or over)	Mice, rats, etc.
Sea snakes	Fish and small eels.

Table 2: Foods usually suitable for most captive herptiles

184

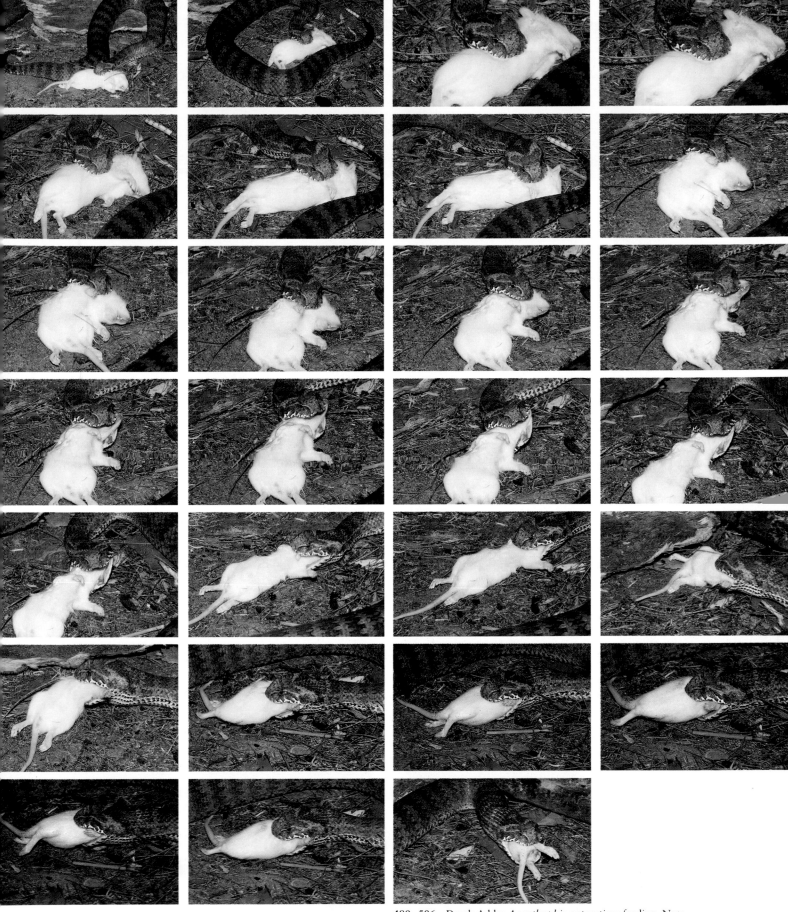

480–506 Death Adder *Acanthophis antarcticus* feeding. Note how the snake positions the food item by moving its jaws over the rodent to get the head into its mouth before beginning to swallow.

185

Force-feeding of herptiles is rarely necessary. It usually involves opening the mouth and gently massaging the food down the throat and into the oesophagus, where the herptile usually begins to finish off the feeding process. Usually a blunt stick-like tool is used to help force the food into the herptile's mouth. 'Tube feeding' involves the placing of a tube, through which food or medicine may be pumped, down the throat and into the stomach. Force-feeding should only be used as a last resort for anorexia, or to administer medicine.

Obesity is rarely a problem in herptiles; however, some larger monitors and other herptiles are prone to becoming grossly obese. For a number of reasons, obesity should be avoided, although it is better to have an obese herptile than an undernourished one.

Food Supplies

For those species which feed on meat and fruit, obtaining food rarely presents problems, but obtaining live food for snakes and insect-feeding lizards and frogs can sometimes pose problems.

Insects may be obtained by several means. By leaving meat out to rot, maggots (fly larvae) will rapidly be attracted in large numbers, and these can be used for food. However, many people have a strong dislike for these, and take to breeding certain types of insect in captivity.

Flour Beetles *Tenebrio molitor* (mealworms) are the most popular insects bred by herpetologists. These insects appear to lack a distinct breeding season and a large number can be maintained relatively easily throughout the year. Mealworms should be maintained in a well-ventilated but sealed container, such as an old ice-cream container, filled with oats interspersed with a few layers of cloth. The small yellow mealworms grow rapidly, pupate, turn into black beetles, lay their eggs and die shortly afterwards. Occasionally pieces of moist vegetable matter should be added to provide moisture for the mealworms. One should avoid removing material from the container as it might contain unhatched eggs. Mealworms or beetles can be removed from the colony as needed. It is advisable to maintain several colonies of more than fifty mealworms at any given time, so that one does not overly deplete one colony leading to later food supply problems.

For snakes which feed on skinks, the only way to obtain food is simply to catch it yourself. One must be careful not to keep more snakes than one can obtain food for. As a general rule, most lizard-eaters eat an average of three of appropriate size per week.

Although many snakes will feed on frogs, frogs should be avoided when possible as they are often an intermediate host for a number of internal parasites which are harmful to reptiles. Fortunately there are only a few species which appear to feed exclusively on frogs.

Pythons and most other medium-sized to large snakes will readily take mice and rats. These food animals can be bred by the keeper without too much difficulty. It is important to keep them well away from the captive snakes, as their constant smell tends to put snakes off their food. Surpluses of mice, rats and similar food should be stored in the freezer for possible use at a later date. When killing food for storage one should not use any chemicals or gasses that may be toxic to the reptile. To thaw out frozen food rapidly, a microwave oven is often useful.

Although most snakes only take live food in the wild, they will in captivity learn to eat pre-killed food. Once a snake is accustomed to eating pre-killed food, it is more desirable for a number of reasons to feed them exclusively on pre-killed food. Reasons include:

1   The snake won't be at risk from attack by the food item.

2   The feeding of the snake can be regulated more easily.

3   One won't have to dismantle or interfere with a cage in order to remove uneaten food.

4   If one has a large stock of frozen snake food, one can more easily guarantee a permanent food supply for all snakes.

For several years I held a large number of snakes of varying types. All were fed exclusively on dead food.

Initial stocks of mice and rats can be obtained from pet shops. Day-old chickens which may be used for snake food can be obtained in large numbers from chicken hatcheries either free or for a nominal price. When given a choice of using birds or mammals for snake food, always use mammals, as the nutritional value per gram of body weight appears to be greater.

A number of methods are used to induce newly-captive snakes (and agamids) to eat dead food. The best method is probably to give the reptile the opportunity to eat pre-killed food immediately after it has just eaten live food, while the reptile is obviously still very hungry. Using long forceps one may be able to hold the dead food item in front of the reptile and wiggle it. The movements will usually cause the reptile to seize the item. Forceps are necessary as the reptile will otherwise attack the handler's hands instead of the food.

When snakes are large enough to take even small mice, for a number of reasons mice are the most desirable food to use. However, often people have trouble converting skink-feeders to mice. A method I used successfully was the following. The mouse would be tied with string to the dead skink and fed to the snake. The snake would strike at and commence to eat the skink. The attacked mouse would also be swallowed. After doing this, one or more times, the snake should become used to taking mice only.

The amount of food eaten over the long term by given reptiles or frogs will vary considerably depending on several factors, including; age, size, average temperature/metabolic rate, parasite load being carried and any ailments present.

Further Useful Points

Reptiles and frogs are extremely clean animals, and their cages should be kept spotlessly clean and free of odour at all times. All faeces, shed skins and other waste materials should be removed from the cage as soon as deposited. Nozzle-style vacuum cleaners are often useful for cleaning out cages. Toxic disinfectants should not be used, as they will harm the herptiles.

When keeping an unfamiliar species, it is important to find out as much about it and its keeping as possible. This can be done by checking the literature, and speaking with others who keep or have kept the same species (or a related species). In other words, learn from other people's mistakes rather than make your own mistakes at the expense of the herptile.

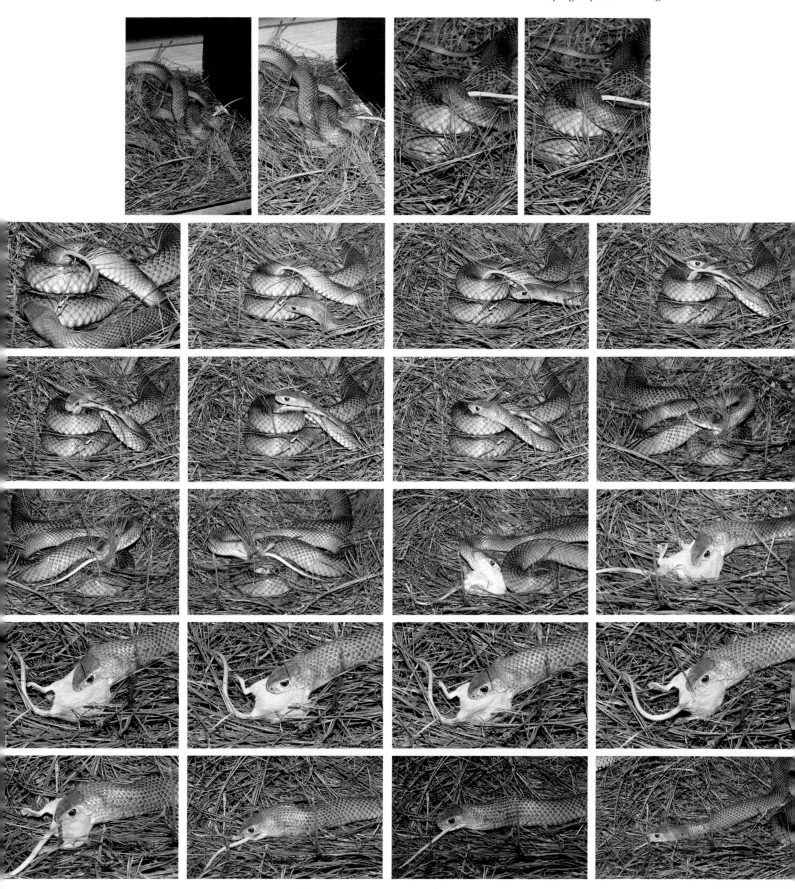

507–530 Eastern Brown Snake *Pseudonaja textilis* feeding. Note how the snake immobilises its prey by constriction as well as relying on the effects of its venom.

A good idea of how to keep herptiles can be gained from seeing how other people or zoos keep their herptiles. Unfortunately in Australia most reptile and frog keepers do not have adequate facilities or take proper care of their herptiles. However, this does not excuse anyone from failing to care for herptiles properly.

When herptiles live for several years and breed successfully, one can usually assume that the herptiles are being kept properly.

It is important to monitor (although not necessarily handle) captive herptiles regularly so that potential problems can be pre-empted. Although handling herptiles generally poses no problems either for the herptile or for its keeper, a general recommendation is to keep handling of specimens to a minimum.

I have always believed in taking accurate and detailed notes on reptiles and frogs held in captivity. Because little is known about the majority of species, notes about captives are vital in finding out more about them and their biology. More importantly, however, by keeping detailed records of specimens held, one can more easily detect potential problems before they become serious. Although I have conducted some of the most detailed captive studies on Australian reptiles to date, there are still minimum records that need to be kept for all captive herptiles by their keepers. These include the following for each specimen:

1 Species (and some identification code or name for the particular specimen).

2 Source obtained from and/or original source in wild.

3 Details (recorded) of the housing of the specimen, including climatic information (necessary for the identification of potential housing problems or improvements).

4 Accurate growth records (by taking measurements).

5 Feeding records.

6 Sloughing records for snakes.

7 All 'unusual' behaviour recorded.

8 Other miscellaneous information.

The records on each herptile should be stored in a safe place close to the herptile in question, so that they can be referred to as necessary. The records must be held for at least the life of the herptile, and preferably indefinitely. The design of the recording system should allow for additions to be made over several years. The information gained from keeping records enables one to make decisions based on facts, rather than taking stabs in the dark when planning for the welfare of the herptile specimens.

New herptiles should be kept separate in a quarantine station and any ailments fully treated before joining other specimens in a collection. This is to prevent the possibility of introduced diseases wiping out a collection. One should systematically remove any internal parasites present, even if they do not appear to be causing any problems.

Keeping herptiles is not as simple as keeping other pets like dogs and cats. Before deciding to keep herptiles one should take these facts into consideration.

CAPTIVE BREEDING

JUSTIFIABLY SO, THE pinnacle of success in keeping herptiles in captivity is breeding them. As herptiles become scarcer in the wild, and more research is carried out, captive breeding will become more important.

Sexing

In order to breed herptiles one must have specimens of both sexes. Frogs, testudines and crocodiles can be difficult to sex but squamates are usually fairly easy.

Sometimes adult frogs of different sexes are of different sizes or slightly different morphology. In some species the forearms are thicker in the males, and these may even possess spikes, which are used to grip the female during amplexus.

Excluding long-necked species, most Australian turtle and tortoise species can be sexed by the presence of the male's distinctly longer tail. In long-necked species it is common for males to have an indented plastron, useful for mounting females, but this is not a general rule.

Male crocodiles have slightly different morphological features to females, but can only be sexed by people who are highly familiar with crocodiles.

All squamates can be sexed by probing. This technique involves the placing of a blunt probe into the base of the tail. In females the probe rarely travels far, as it is stopped by the flesh within the tail, usually within three scales of the vent. In males the probe will go further into the tail, moving into a hemipene, and generally travelling further than three scales. Probing is a very delicate operation, and should not be attempted by anyone who has not seen it performed by someone familiar with the procedure. 'Misprobing' can make a reptile sterile.

Male geckoes can be distinguished by the presence of a swelling at the anal region. Male dragons and pygopodids usually possess pre-anal and/or femoral pores on the underside of the hind legs.

Some male pygopodids possess pelvic spurs in addition to their hind limb rudiments.

In skinks and agamids, males are often of slightly different colour from females, often with flushes of red or other bright colours.

531 Scalyfoot *Pygopus lepidopodus*, ventral surface, showing pre-anal pores (above vent) and rudimentary hind limbs (Mount Glorious, QLD).

532 Scalyfoot *Pygopus lepidopdus* with rudimentary hind limb raised to reveal pelvic spur.

533 Sexing an Olive Python *Bothrochilus olivaceus* by probing. Both pelvic spurs are visible. (Pine Creek, NT).

534 Unusually large pelvic spurs in a male Green Python *Chondropython viridis* (origin of specimen unknown).

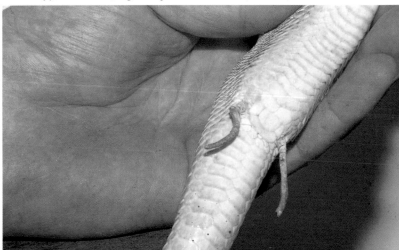

535   Giant Burrowing Frogs *Helioporus australiacus* in amplexus (West Head, NSW).

536   Womas *Aspidites ramsayi* copulating. The male is the larger specimen. (Male from The Tits, WA, female from Tea Tree, NT).

Many large monitors will evert their hemipenes when handled.

In some snakes, males have distinctly larger tails than females. However, unless one is particularly familiar with a given species, probing is usually the only possible way to sex a snake.

Initiating Breeding

Usually herptiles that are kept properly will breed as a matter of course. To date few attempts have been made to breed Australian frogs in captivity. Tortoises and crocodilians kept in captivity for a number of years produce eggs largely as a matter of course.

Snakes and lizards, however, seem to be the hardest reptiles to breed in captivity. It is believed that both males and females have sexual cycles. When their cycles coincide and mating takes place, breeding occurs.

If reptiles kept indoors experience no temperature variation during the year, they often fail to breed. Some cooling is required to initiate spermatogenesis in males and ovulation in females. Most breeders cool their snakes and lizards for six to twelve weeks before the anticipated

breeding. In some species, breeding cycles occur up to nine months after the initial cooling, meaning that production of offspring may take up to eighteen months from a given cooling period.

Cooling involves dropping the reptile's temperature so that it is too cold to feed, and it goes into a relatively inactive state. No reptile should be cooled to a point where it may be harmed. 15–20° C is usually sufficient to initiate breeding activity.

To initiate mating, separation of the sexes is usually advised. When isolated males cease feeding and pace their cages, they are usually interested in mating. By placing them with a receptive female at this time, successful copulation will usually occur. As reptiles are most sexually active during periods of falling air pressure, often associated with fronts, these are the periods that separated sexes should be reintroduced to each other.

Some reptile breeders vary photoperiod (daylight hours) in a bid to initiate breeding activity.

Gravid (pregnant) reptiles must be treated with extreme care, as the wrong set of temperature or other conditions can lead to death of unlaid eggs or unborn young. Gravid reptiles will abort if kept too warm or cool.

Incubating Reptile Eggs

When eggs are laid, they should be removed from the cage. The exceptions to this may be in the case of pythons which will attempt to incubate the eggs themselves, or reptiles which lay eggs in specially chosen sites in outdoor pits. Eggs should never be shaken or even turned over as this will kill the developing embryo. Most people mark the top of eggs with a cross, or preferably a number, with a pen, so that the orientation of the egg can be maintained if inspection of the egg is required during incubation.

Most eggs are best incubated at between 25° and 30° C. They should not be exposed to temperature fluctuations of more than about 1° C.

Eggs found adhered to one another should usually not be separated from one another, as this may cause breakage.

A number of different media can be used to incubate eggs. Eggs are best incubated in a container, such as almost any type of box, which have in place a medium such as sand, vermiculite, peat moss or similar. The eggs may be buried within the medium, although it is best to have the eggs sitting half buried on the surface, so that they can be most easily observed but won't easily move.

Eggs should be placed apart from one another, so that if one egg becomes diseased, it will not jeopardise the other eggs. In order to minimise risks to the eggs, sterile incubation media are preferable. When one egg is obviously infertile, or for some reason not going to hatch, it should be removed from the incubator containing the other, still fertile eggs. Infertile eggs are often distinguished by being an unusual colour, having an unusual shape or size, feeling unusually hard or soft, or having an unusual smell. Before one decides an egg is infertile one should be familiar with what a normal egg for that species looks like.

The humidity at which eggs are kept should be high, usually 80–100 per cent. No eggs should actually have water on them in normal circumstances.

If eggs appear to sweat it is probably because moisture levels are too high, and this must be corrected. Eggs often shrivel up slightly during incubation. If eggs start to shrivel in the early stages of incubation it may be because insufficient moisture is present. If this is the case, it may be acceptable to spray a small amount of water directly on to the egg.

Fungus might be found on an egg. This will usually be on the underside. Should this occur one should wipe the underside of the egg clean of all surface fungus, change the incubation medium, and keep the affected area bone dry. A very dilute anti-fungal solution may be applied to the affected areas of the egg, although extreme care is required here.

To monitor development of an egg, one can hold it against a very bright light and usually see the developing embryo within.

After the first egg has hatched, the others should be watched closely to make sure that hatchlings have no trouble escaping from the egg. If unable to escape they may suffocate. If there is any doubt about hatchlings emerging from their eggs, then it is advisable to slit the tops of those eggs. The young should not, however, be removed from the eggs. It is common for a young reptile to remain within the egg for some time after rupturing the shell with its egg tooth.

537 Gravid Ant-hill Python *Bothrochilus perthensis* four days before laying. The two eggs are clearly visible. The snakes were observed copulating three months earlier. (Shay Gap, WA).

538 Same Ant-hill Python incubating newly laid eggs. Note how the female completely covers the eggs.

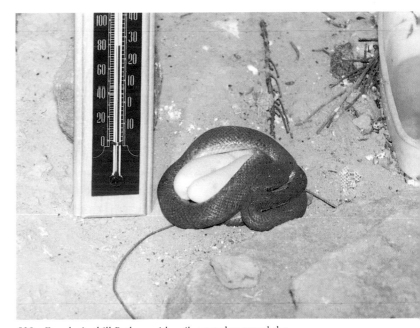

539 Female Anthill Python with coils parted to reveal the unusually large size of the two eggs.

540 Incubation set up of two Anthill python eggs (removed from female by author) which take about 60–70 days to hatch. Note the marking of each egg, with a number and a line, so that regular inspections can be made without altering the orientation of the eggs.

541 Blue-bellied Black Snakes *Pseudechis guttatus* copulating (QLD).

542 Tiger Snakes *Notechis scutatus* copulating (VIC).

543 Death Adders *Acanthophis antarcticus* copulating (NSW).

544 Gravid female Death Adder *Acanthophis antarcticus* (NSW).

546–548 Emergence of newborn captive bred Death Adder from sac (NSW).

545 Same Death Adder giving birth to live young in March 1984. Note the grey colour of all offspring. This means that the successful mating was with a grey Death Adder.

549 Twenty captive bred Death Adders (from three females) (NSW).

550 Newborn Death Adder *Acanthophis antarcticus* (NSW).

551–553 Death Adders *Acanthophis antarcticus* mating and feeding simultaneously (NSW).

554–557 Desert Death Adders *Acanthophis pyrrhus* copulating and feeding simultaneously (WA).

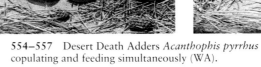

558 Male Desert Death Adder *Acanthophis pyrrhus* attempting to mate with a male Death Adder *Acanthophis antarcticus* (WA and NSW).

AILMENTS, DISEASES, PARASITES AND THEIR TREATMENT

MOST HERPTILE AILMENTS are a result of poor conditions in captivity. The old saying, 'Prevention is better than cure', is very applicable to captive herptiles. Treatment and/or medicines can become expensive for some types of ailment.

When a herptile is diagnosed as having an ailment, the key to success is to act immediately. Failure to do so will usually result in the death of the herptile. In most cases, sick reptiles and frogs take a long time to die. Some people are lulled into a false sense of security when an apparently ill herptile fails to die. Many ailments, when they appear, don't appear to affect the health of the herptile, but must be acted on *before* they reach a more advanced (usually fatal) stage.

Virtually all ailments are highly contagious and affected herptiles should always be isolated from other captive specimens.

When in doubt as to how to treat an illness, one should seek the advice of a qualified person such as a veterinary surgeon, or preferably an experienced herpetologist. The average suburban vet sees very few reptile diseases, however, so more than one opinion may be required.

Treatments to more common ailments are given below. Dosages required depend on the potency of the treatment being used and the size and robustness of the herptile in question.

Should a herptile die of 'unknown causes' a post mortem should be carried out by someone experienced to find the cause of death.

There are various ways to administer oral medicines to reptiles and frogs. The medicine may be inserted into the food or food animal and then fed to the herptile. For example, a tablet can be placed under the epidermal layer of a mouse and then fed to a snake. The medicine can be fed or force-fed directly to the herptile, although this is usually difficult. Often one may tube-feed the herptile. This involves placing a tube into the herptile's mouth, leading to the stomach, and then putting the medicine or food through the tube.

Injectable medicines must be injected in a fleshy part of the body, such as on the back, adjacent to the spine or in the base of the tail. If injecting into the base of the tail in male specimens, one should be careful not to inject too close to the hemipenes.

Leeches — Frogs and Tortoises

Leeches commonly act as parasites to freshwater tortoises and frogs, and can kill both, so they must be avoided at all costs.

If leeches are found in a cage, the insides of the cage should be thoroughly cleaned out, and the leeches removed. Simultaneously the specimen should be cleaned of any attached leeches and then kept in a leech-free cage. By immersing the specimen in salt water external leeches will drop off, although one should take some care with frogs, which are themselves susceptible to death from exposure to salt water. The inside of the mouth must also be checked for leeches. For tortoises the addition of a little salt water in outdoor cages will generally keep leeches away from the water, where they usually attack the tortoises.

Anorexia — All Herptiles

Most herptiles will abstain from feeding at particular times of the year, usually in response to breeding or hibernation activity. Their health will usually not be impaired by this abstinence. However, when a herptile fails to eat to the point where its health is obviously in peril, this abstinence must be treated as an ailment.

Failure to feed normally is usually a response to incorrect keeping of the herptile. This may be due to faulty cage design, temperatures, etc, or coinhabitants dominating the herptile that will not eat. Anorexia may also be due to the effects of some other ailment, possibly undiagnosed.

If no apparent reason can be found for the anorexia, force-feeding may be necessary. When force-feeding it is advisable not to use large food items, as the herptile may have difficulty in digestion. Herptiles can be kept alive by force-feeding for indefinite periods. Relatively thin herptiles can be made fairly obese by force-feeding if desired. Force-feeding small amounts of food on a regular basis is most likely to act as a stimulant to restore normal feeding behaviour. The administering of some vitamins and minerals may also act to stimulate appetite.

Fungus — Tortoises and Frogs

Indicated by white or grey patches, fungus can spread rapidly to cover the entire body. The affected area should be painted with codeine or mercurochrome and/or a 60 per cent diluted alcohol solution. The procedure should be repeated every twelve hours. Aquarium fungicide should be added to all cage water during and shortly after treatment, and as necessary to prevent infection recurring.

Tortoises are best kept relatively dry and allowed to bask in as much sun as possible during treatment.

Scale Rot — Snakes and Lizards

Scale rot is usually a fungal condition and usually occurs when reptiles are kept in surroundings which are too cold and/or moist. When affected scales, usually the ventral ones, appear to die and eventually fall away from the body. The affected areas should be dried, cleaned and painted with a drying solution such as mercurochrome. This process should be repeated as necessary.

The reptile should also have its surroundings altered so that the condition does not recur. Scale rot can sometimes take quite a while to heal.

Soft Shell — Tortoises

Rarely a problem in itself, a soft shell can make a tortoise highly susceptible to other ailments, and must therefore be prevented. Sunlight and vitamin D help to harden the shall. Calcium in various forms may be added to the pond water, and calcium-rich food such as meat and fish should be offered.

Respiratory Infections — Reptiles

Respiratory infections are commonly diagnosed when the reptile suffers from nasal congestion, discharges from the mouth or nose, forks of tongue sticking together, listless behaviour, etc. Also reptiles will usually stop eating when afflicted with respiratory infections.

These infections are usually caused by excessively cold or moist cage environments. Treatment is by keeping the reptile relatively warm and by administering sulphur drugs such as sulphdimazine tablets daily. When possible it is desirable to remove excess mucus and other blockages.

Body Infections — Reptiles and Frogs

Body infections are usually a result of dirty or excessively damp cage conditions or abrasive surfaces (particularly for tortoises). Infections can be treated with a variety of antibiotics, including Neosporin, Terramycin, etc.

The affected surface should be cleaned of dead tissue with a swab, and the antibiotic powder, ointment or cream should be then applied to the surface, usually every twelve hours. When it is difficult to remove dead skin, scales, etc., from a reptile, soaking it in lukewarm water for a few hours should make removal easy. The reptile or frog should be held in a cage where debris won't enter the infected area and cause complications.

Mouth Rot (Canker) — Most Reptiles

Most common in Pythons (Boidae), this disease affects the gum tissue of the mouth, and is often associated with respiratory infections (see above). It is caused by a bacterium, usually *Aeromonas* or *Pseudomonas*, which apparently usually inhabit Pythons' mouths without causing problems. Treatment of this disease is tricky and requires persistence.

The affected areas of the mouth should be swabbed with diluted antiseptic such as Dettol or Listerine every six hours. The dilution should be the same as a human would use for a mouthwash. Twice daily an antibiotic (powder is usually best) should be applied to affected areas immediately after being swabbed with antiseptic. In all but the mildest cases intra-muscular injections of antibiotic should also be used on a 24- or 48-hour basis *in addition* to the other treatment.

When all visible signs of the disease are gone, the mouth should be swabbed twice daily with diluted antiseptic for at least ten days.

More so than most other ailments, mouth rot has a habit of recurring after treatment, so recently affected reptiles should be monitored particularly closely and kept in isolation for a few months after the disease has apparently gone.

Mites — Snakes and Lizards

Mites are without doubt one of the greatest killers of captive reptiles. This is a tragedy as mites are so easy to control. These tiny arachnids live under scales and congregate around the eyes and similar places. The types that kill reptiles are usually red or more commonly brown in colour. Mites breed very rapidly (30–90-day lifecycle in most cases), and when present in large numbers kill the reptile by blood poisoning. Reptiles infested with mites typically have raised scales, and the white droppings from these mites are usually highly visible on the scales as white dots. In particularly bad cases mites will be seen walking over the reptile.

Infected reptiles commonly soak themselves in a bid to drown the mites.

In order to kill off mites both reptile and cage must be cleaned of mites. Shelltox Peststrips, with vapona as the active ingredient, are the safest and most effective means of combating mites. Infected reptiles should be placed in a different clean cage with an appropriate amount of strip. The reptile does not usually need to be shielded from the strip in any way, as they have extremely strong resistance to these strips. The cage should have a strong, but not overpowering, smell of pest strip. Dead mites should be visible in the cage. Once the reptile is apparently mite-free, it must be maintained in the new cage with the strip in order to kill off newly-hatching mites. The reptile must be assumed to be mite-infected for up to twelve weeks after initial treatment, although the amount of pest strip present may be reduced after forty-eight hours.

A very few reptiles may show some reaction to the pest strips. In these cases the concentration is too great and

559 Death Adder *Acanthophis antarcticus* with severe mite infestation. The white spots are mite droppings. Specimen courtesy Royal Melbourne Zoo, VIC.

should be reduced. Pest strips should be kept away from water, because if they drip into the water, and the reptile drinks it, it may be poisoned and die.

The cage in which the reptile was originally kept must be disinfected totally of mites. The easiest way to do this is to place pest strip within the cage so that the smell is overpowering. The cage should be left like this for twelve weeks, when the strips may be removed. When the smell of vapona has subsided reptiles may be reintroduced.

Pest strips can be kept in reptile cages to prevent mites without affecting the reptiles in any way.

At all stages of mite treatment the affected reptiles should be monitored very closely.

Drie Die dust is a powerful desiccant, which is also effective against mites. Reptiles are known to have eaten food with Drie Die attached without apparent problems. Neguvon powder when mixed with water is also effective in disinfecting reptiles and cages, although it can be toxic to reptiles.

The former practice of applying oil to reptiles in a bid to suffocate mites should be avoided as this will also often kill the reptiles.

reptiles free of these parasites. Ticks kill reptiles by injection of poison and excessive taking of blood.

Large ticks are removed by tweezers. When removing ticks it is important to make sure that the head and associated mouthparts are also removed. Failure to do so could result in further complications. Smaller ticks may be removed using the same methods used for mites.

There is a minor risk of infections occurring at points where ticks are removed.

Body Blisters — Snakes and Lizards

Mainly occurring in snakes, these blisters usually arise from excessively damp conditions and generally poor health of the reptile. If these factors are controlled the blisters should subside. For bad cases, the reptile should be bathed in dilute antibiotic solution, such as terramycin poultry formula, for about two hours once every three days, about three times.

Ticks — Reptiles

Most reptiles have a strong degree of immunity to ticks. However, ticks reproduce rapidly and can kill reptiles when present in large numbers, so it is important to keep

Internal Parasites — All Reptiles and Frogs

Internal parasites lead to a host of problems for herptiles

and are often hard to isolate and treat. Nematodes (Nematoda), Tapeworms (Cestoda) and Flukes (Trematoda) all live on ingested food and when they multiply sufficiently can kill the reptile. Eggs and adult specimens are usually passed in the faeces, so these should be regularly checked. Pet worming tablets containing Piperizine will remove most internal parasites, and should be administered at about the same dose rate as for mammals.

If a herptile is feeding voraciously and not gaining weight, it is likely that it is being affected by some type of parasite.

An enlarged or swollen heart is also usually a sign of infestation by some kind of internal parasite.

'Skinworms' are most common in frog-eating snakes and are often visible externally as cysts under the scales of the snake. These external worms may be removed by making a small slit in the skin between the scales and pulling the worms out with tweezers. Other skinworms present inside the digestive tract must be removed with oral treatment (if necessary). Skinworms disfigure the appearance of the reptile but are rarely harmful. Dead skinworms remain encysted under the snake's skin and harden, being harder to remove than the living worms.

Blood flukes are often hard to detect, and often harder to treat.

Internal parasites may produce secretions which prevent the digestive system from operating normally and lead to large calcerous deposits within the intestine, which may block it and kill the reptile.

Larger parasites, such as Tapeworms and Pentastomids (Lung worms) (Pentastomida), can cause blockages in the respiratory system and elsewhere, leading to rapid and often unexpected death. Tapeworms of more than 5 metres are known.

Once a parasitic worm has been positively identified, it is relatively easy to locate the relevant drugs, etc. necessary to treat the infestation.

To identify and treat most types of internal parasite it is usually best to enlist the assistance of a pathologist or relevant laboratory.

560 Skinworm, removed with tweezers from a Golden-crowned Snake *Cacophis squamulosus* from West Head, NSW.

561 Broad-banded Sand Swimmer *Eremiascincus richardsoni*. The swelling in the throat region is caused by a Pentastomid worm infection. Although apparently harmless to this species, some reptiles, notably Death Adders *Acanthophis* spp., are known to suffer adversely and even die when infected with these parasites. This specimen came from northern South Australia

Gastroenteritis — Reptiles

Caused by protozoans of the genus *Entamoeba*, gastroenteritis can kill reptiles within forty-eight hours of onset. *Entamoeba* infestations often occur through contaminated drinking water and a reptile may be infested for up to twelve months before coming down with gastroenteritis. It appears that reptiles kept constantly at more than 27°C are usually immune to infection by *Entamoeba*. Gastroenteritis occurs when the *Entamoeba* multiply extremely rapidly within the lower intestine of the reptile. This is usually during times of temperature stress (spring and autumn). The *Entamoeba* produce massive amounts of gasses and the reptile's body may swell considerably at the rear end. Advanced cases paralyse the rear end completely and these specimens rarely survive, even with treatment. Affected reptiles have severe diarrhoea and pass mushy, smelly faeces.

Treatment involves the oral administration of anti-protozoal drugs, such as Iodochlororhydroxyquinoline (Vioform and Flagyl), usually once or twice.

When one reptile in a cage shows symptoms of *Entamoeba*, all reptiles in the same cage must be treated for infection.

PRESERVING SPECIMENS

FOR A NUMBER of reasons it is often necessary to preserve live or dead reptiles and frogs. If a dead herptile is to be dissected within a few months of being killed, it is sometimes best kept frozen. When other preservatives are not available, a freezer is always an excellent temporary storage place.

If live specimens are to be killed to be preserved (usually only for museum and scientific collections), then freezing of specimens is the most humane and painless way to do so. The metabolic rate is slowed down gradually until it stops and death occurs. Various anaesthetics may be used to kill specimens, but they are not as desirable as killing by freezing.

The best solution for preserving herptiles is 60–75 per cent ethyl alcohol. The lesser concentrations should be used for softer-bodied herptiles, while higher concentrations should be used for larger and harder-bodied herptiles.

A solution of 7–10 per cent formalin is also excellent for preserving specimens. Frogs, tadpoles and frogs' eggs are best preserved in 8 per cent formalin with small amounts of calcium chloride and cobalt nitrate added. (Tadpoles and frogs' eggs should never be preserved in alcohol-based solutions.)

Methylated spirits and other alcohol-based solutions, including high alcohol liquor, make good preservatives, and are readily obtainable even in remote areas.

It is often advisable to inject specimens with preservative when preserving them, to avoid decay before the preservative seeps through to the internal parts. Fixing specimens into desirable positions can be done by injecting them with preservative (formalin is best for fixing), and then positioning the body into the desired position before it stiffens. Hemipenes of snakes and lizards can usually be everted by injecting preservative into the base of the tail behind the hemipenes.

Addition of 10 per cent glycerine to preservatives may help maintain softness of specimens.

Specimens should be stored in screw-capped plastic or glass jars, with as little air within as is possible. The seal should be perfect to prevent evaporation of preservative, and metal lids should be avoided as they will corrode and eventually leak.

Detailed records should be kept in the same way as for live reptiles. For each specimen one should record the following:

1 Species.

2 Date.

3 Locality.

4 Person who obtained specimen.

5 Other significant data.

Specially made labels for preserved specimens are available. These contain a file number and may be tied to the specimen within the container. The specimen number correlates with notes kept close to the preserved specimen. One should make sure that labels and writing attached to specimens is not adversely affected by the preservative. Labels made of rag-based paper, and written in pencil or indian ink, are usually effective.

A WORD OF WARNING

IN ALL AUSTRALIAN states there are varying degrees of so-called protective legislation for herptiles. The legislation only covers capture of herptiles and herptiles held in captivity (dead specimens included), and is administered by state wildlife authorities. In theory, any individual who wants to take, kill or do research on herptiles must usually contact the relevant state authority and obtain permits. Failure to do so may be illegal. However, in some states, particularly New South Wales and Queensland, otherwise law-abiding citizens prefer to hold reptiles and frogs without the knowledge of wildlife authorities. This is because of actions taken by the authorities against reptile licence-holders, including break-ins and theft of specimens. The roaring (illegal) trade in reptile specimens has resulted in corruption within statutory bodies appointed to protect wildlife, and consequently the probability of break-ins, theft and harassment make any involvement in the keeping of herptiles a risky undertaking.

Although I cannot condone holding protected fauna without permits, harassment suffered by myself and colleagues in the period 1974–87 probably would not have occurred if we had held reptiles without permits during that time.

Before making contact with state fauna authorities in relation to obtaining permits for herptiles one should:

1 Check whether a permit is really required to take, kill or keep the particular specimens, and

2 If a permit is legally necessary, one should seriously consider changing one's course of action.

All state fauna authorities have a habit of regularly changing laws, so it is important when investigating current regulations to bear this in mind.

PHOTOGRAPHING REPTILES AND FROGS

PHOTOGRAPHING REPTILES AND frogs is essential to those who take a scientific interest in herptiles. It can also be a rewarding pastime and should be encouraged. Along with good records, photos can be used to identify a given reptile or frog specimen. This is particularly important to those who hold captive specimens and might need to identify particular specimens when stolen.

The quality of photos taken depends not only on one's skill but also on the quality of one's camera equipment. One can spend literally thousands of dollars on camera equipment.

Through house break-ins, I have lost all my camera equipment twice. Consequently the photos in this book have been taken with a range of different cameras and equipment of varying types and brands.

Equipment

Camera

The piece of equipment fundamental to photography is the camera. A single lens reflex (SLR) camera capable of accepting interchanging lenses is mandatory. Although various film size formats are available, 35 mm is usually most desirable on the grounds of price, availability of special lenses, and the quality of photographs produced. Larger film size formats are usually prohibitively expensive. Most professional photographers use 35-mm cameras.

Price is usually a good indication of quality, so one should spend as much as possible on a camera and other photographic equipment. By shopping around and/or purchasing duty free, much money can be saved. Well-known quality brands such as Nikon, Canon, Pentax, Olympus, Minolta and Fujica should be used. All these are essentially the same except for features on particular models and lens-mounting systems. As a flash is used for about 90 per cent of reptile and frog photography gimmicks on the camera are of little importance.

Lenses

Most reptile photography involves close-up photography and therefore special lenses and other equipment are needed. Methods used for close-up photography are varied and include the following:

1 'Macro' lenses, 50 ml and similar.

2 Variable focal length lenses with 'Macro'.

3 Diopter lenses.

4 Teleconverters.

5 Extension tubes.

6 Bellows.

7 Lens reversing rings.

Lenses bought must be compatible with the camera. This usually means that they must have the same mounting system or an especially made mount.

The lenses used must be of top quality as a substandard lens will always produce substandard photographs, regardless of the photographer. Things to check for in a lens are clarity of image, and image distortion when close up.

Of the above-mentioned methods of getting close up, the most popular with herpetologists is a standard focal length macro lens. This enables one touch focusing at all distances from very close to infinity. The relatively small size and lightweight nature of this lens allows for maximum manoeuvrability around a subject at close quarters.

I rarely use a standard focal length macro lens or similar when taking photographs of reptiles and frogs. I prefer to use a variable focal length lens with macro (usually 70–210 mm) mounted on a tripod to take the majority of reptile photos. The lens style enables me to frame the photograph adequately, and by moving the tripod if necessary, manoeuvrability is not lost. A tripod is essential for this and other heavy lenses in order to prevent unwanted camera shake. In reality I virtually always have my camera tripod-mounted when photographing reptiles and frogs, regardless of lenses used.

Diopter lenses, extension tubes, bellows and reversing rings are regarded as being too fiddly for most reptile and frog photographers.

To get extremely close (as in, say a head photo of a small lizard) several methods may be used. I use seven element two-times teleconverters (up to three in series), as well as a macro lens (about 50 ml) to get ultra close focus. The above apparatus, like all extreme close-up set-ups, does give substantial image distortion, but is better than alternative set-ups.

Because this set-up of lenses with the teleconverters uses so much light (128 times that of the naked lens) it is often necessary to provide powerful lighting just to focus the subject. I use a 2000-watt floodlight, tripod-mounted at close range, to provide focusing light. I may use one or two teleconverters in series for intermediate close-up photography.

Teleconverters come in a variety of forms, usually specified by the number of lens elements. It is critical to get at least seven element teleconverters for reptile and frog photography.

To get even closer, microscope attachments are available. These are relatively cheap (about the cost of six rolls of slide film).

All lenses bought should be covered at all times with a filter. Not only do these remove some unwanted light rays that may spoil photos, but also they are cheap and give scratch protection to the lense, which is usually very expensive to replace.

Light Source

Focusing the subject is usually no problem. If lack of light causes focusing problems a floodlight, photographic lamp or similar may be used. I prefer to use floodlights as they are considerably sturdier and less prone to malfunction, and cheaper too!

For photographing reptiles and frogs the key to success is maximum depth of field (line of focus). To obtain this one must take photos with the lense stopped down to the smallest aperture. (The smaller the aperture, the higher the F-number.) For most reptiles and frogs F-11 is a bare minimum for usable depth of field. Most lenses go as far as F-22 and one should try to use this when taking photos of most reptiles and frogs.

One should remember that the closer one gets to a subject, the less the depth of field and the greater the need for a small aperture.

If relying on natural light, your shutter speed will regulate your aperture (depth of field). Unless using 'fast' film one would rarely be able to attain maximum depth of field at above 1/60 of a second. Below this shutter speed the risk of camera shake and/or subject movement blurring the picture becomes unacceptably high.

When using a flash, the shutter speed is set at a given, specified 'flash sync' speed, usually 1/60, 1/100 or 1/125 of a second. At close range most flashes are considerably more powerful than the brightest sunlight. As one needs as much light as possible to get maximum depth of field I recommend the largest flash possible (guide number 45 upwards).

When taking photos the flash may be mounted either on or adjacent to the camera, or elsewhere depending on requirements. When taking most reptile and frog photographs I have the flash mounted at a predetermined distance from the subject (often on top of the camera), and move the tripod-mounted camera if necessary. Although most herpetologists use automatic settings on their flashes, I prefer to use maximum, full power manual setting in order to maximise light output and possible depth of field. When taking extremely close-up photographs with added teleconverters, the flash may be mounted as close as 15 cm from the subject. (In this case getting correct exposure is tricky and requires accurate distance measurements to be taken regardless of flash mode used.)

A flash may be used to fill in shadows caused by sunlight, or two flashes may be used in combination when photographing subjects, with the aim of minimising harsh shadows. A double shadow effect may occur, and I find this unappealing. In all photos harsh shadows or 'black spots' should be minimised. In most cases this is best accomplished by having the flash-head mounted above the camera and subject, and between them, or to have the subject facing the flash head if the flash is side-mounted.

Powerful studio lights may be used to provide light for photographing reptiles and frogs. In general these are undesirable as they still fail to provide sufficient light, are highly prone to mechanical failure, and special blue filters are usually required.

Film

'Film speed' is a term used to describe the light sensitivity of a film. The higher the rating (film speed), the less light is required for taking photos. For each DIN increase or ASA doubling one can double shutter speed or gain one more F-stop. As depth of field is all important to reptile and frog photography, fast film is usually more desirable. The trade-off is that fast film produces grainier results than slow film. ('Grain' shows up as small deposits or spots of silver iodide when the photo is enlarged.)

The total trade-off becomes the need for depth of field, combined with the lens's light requirements, *versus* the power of the flash. Film speed chosen should reflect the need to obtain maximum depth of field, but should not be any faster than necessary. For example, using a 50-ml macro lens and a guide number 45 flash at one metre for most photos, 64–100 ASA film would probably be most desirable for herpetological purposes.

Most people prefer colour film to black and white, and today black and white is used mainly for press or scientific publications. Colour slides (positives) give better reproductions than print negatives, although what is finally used should be a matter of personal need and preference.

Tripod

A tripod is one of a number of accessories many reptile and frog photographers use at various times. I depend heavily on them. When taking extreme close-up photographs I will use three tripods: one for the camera, one for the flash and one for the floodlight to provide focusing illumination.

Usually the function of a tripod is to prevent camera shake. For this reason it is important to buy one of rugged, sturdy construction. The tripod should be easy to assemble and dismantle, and pan and tilt smoothly and efficiently.

Other Accessories

Sturdy camera cases are essential to protect photographic equipment, as is useful camera cleaning gear such as lens tissues and air spray cleaners. A number of miscellaneous camera attachments may be used by herpetological photographers. These include infra-red remote flash controls, shutter cords, etc.

562   The author photographing a Scalyfoot *Pygopus lepidopodus*. The photographic equipment shown is similar to that used to take most photographs in this book. The container below the tripod is used to restrain specimens that might otherwise run away while positioning camera equipment before taking a photograph.

Posing the Subject

It usually takes some practice before one can take correctly exposed photos of small animals, with maximum depth of field in all types of conditions.

Reptiles and frogs will not pose for photographers while in the wild. It is effectively impossible to photograph a wild reptile or frog in a 'nice' position without disturbing it. Therefore all specimens to be photographed will need to be caught and held before being posed for photographing.

As far as getting a subject to pose in a desirable position goes, it is fair to say that the beginner is usually easier to please than the expert. A 'good' or 'bad' photo in terms of composure cannot be strictly defined as it is a matter of personal opinion. However, most herpetologists tend to agree on a number of features worth having in most photos of reptiles and frogs.

When possible, the background should reflect the natural environment from where the specimen originated. There is no point in photographing a reptile from coastal areas on red sand and spinifex, or *vice versa*. It is fairly easy to construct a small stage on which to photograph specimens, with bits and pieces of logs, rocks, gravel, sand or whatever, that will look like a natural setting. By photographing down on to the subject one only needs an area of 'natural setting' slightly greater than that occupied by the posing specimen itself. It goes without saying that one doesn't want bits of carpet or table top in the corners of the picture. Most of my photos of reptiles and frogs are on the type of stage just described. The subject should be in the middle of the photo, and ideally fill most of the frame.

When taking body shots it is best to have as much of the body showing as possible. The dorsal surface should always be visible and as much lateral surface as possible should also be visible. The head, in particular the eye, is the most important part of the animal and this part should always be in focus. Head shots by themselves are also very important to herptile photographers.

When photographing specimens, important and unique anatomical features should always be visible. Examples include the black spines on the toes of the male Giant Burrowing Frog *Helioporus australiacus*, or the unusual tail of the Death Adder *Acanthophis antarcticus* used in caudal luring. (It is obviously difficult if not impossible to show things such as pelvic spurs in general body photos.)

Long specimens, including snakes, should be coiled as opposed to being in a line, as this allows one to get closer to the subject. A photograph of a large, fully stretched snake will look little different from a piece of rope, with few features being visible.

Getting the subject to sit still is always a problem. Cooling specimens by placing them in a fridge or freezer is a common trick of most herpetological photographers. By slowing down the metabolic rate and movements of the herptiles the risk of the subject escaping while being photographed is reduced. Most cooled specimens are more reluctant to move anyway, often choosing to bask on the stage while being photographed. About a fifth of the reptiles and frogs I photograph are cooled beforehand. This includes most small skinks and highly venomous snakes.

563 Not all herptile photographs have to be scientific. The two Lace Monitors *Varanus varius* pictured were held in captivity for more than twenty years. Both specimens were 'dog tame'.

Agamids are best heated in order to bring out their brightest colours. As they are usually very fast when hot they should be photographed in a sealed room, where they can't escape if they run off the stage.

When cooling or heating reptiles or frogs it is important not to let the specimen's body temperature drop below or rise above those normally experienced in the wild state.

When photographing frogs it is advisable to dip the frogs regularly in water to keep their skin free of debris that will otherwise stick to them. Regular immersion in water also prevents dehydration and skin irritations, and generally makes frogs less jumpy.

Often a specimen can be placed under a container, such as a pot or bucket on a stage, where it will hopefully settle down. Once the specimen stops moving, the container is lifted and the specimen is photographed before it moves away.

The pinnacle of a herpetologist's photographic career is photographing copulating reptiles or frogs or their laying eggs. With the exception of frogs, who are easy to find breeding in the wild, breeding activity in reptiles will usually only be seen and photographed in captive specimens. All breeding activity photographed in this book was photographed in captive collections. (Virtually all were my specimens.)

When breeding, reptiles and frogs are usually too busy to be put off by photographers taking pictures of their activities. Two mating Death Adders *Acanthophis antarcticus* continued copulating after a 2000-watt photographic light dropped on top of them and smashed. (Young were produced about nine months later.)

No aspect of herptile photography should be cruel to specimens. The discomfort to specimens through being handled, cooled, etc, is only minimal and should not discourage one from taking photos of herptiles.

564–571 Painted Burrowing Frog *Neobatrachus sudelli* burrowing. Note how the frog burrows in a backwards motion, using the metartarsal tubercles located on the hind feet. The frog tends to rotate its body when burrowing.

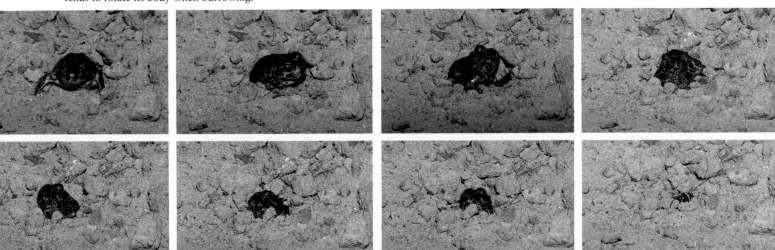

CONSERVATION OF REPTILES AND FROGS

ONE OF THE primary aims of this book is to improve conservation efforts within Australia, particularly with respect to herptiles, which have suffered greatly in the past.

In this discussion of conservation of Australia's herptiles, personal opinions are given, which are based on facts, some of which are not referred to here. It is hoped that interested readers will investigate some of the references cited, so that they may see how my conclusions were reached.

Australia's conservation record is arguably one of the worst in the world, on a population *versus* land mass basis. Australia holds the dubious distinction of being the only nation on earth to have exterminated a reptile through nuclear means. In the 1950s the British were allowed to test the atomic bombs on an offshore Western Australian island, killing off an endemic legless lizard (Pygopodidae).

Reptiles and frogs need not only be conserved for conservation's sake. They also constitute a valuable natural resource. For example, venoms of some species may prove useful in the manufacture of drugs, while many species eliminate pests such as introduced mice, rats, insects, etc.

There are about a thousand species of reptile and frog found in Australia and adjacent seas, including undescribed forms. Although there is an estimated population of about 15 billion herptiles in Australia, many species are rare, have restricted distributions, or both. An annual mortality rate of about 4 billion is replaced by a similar number of new specimens. However, about a quarter of the species known are known to be in a state of either moderate or serious decline. It is these species to which we should be devoting major conservation efforts.

The main threats to herptiles in this country all result from human activities, past and present. In order of importance the main problems facing our herptiles are:

1 Habitat destruction and/or modification.

2 Introduced pest species.

3 So-called 'protective' legislation, 'inconsistent' fauna authority officials and smuggling rackets.

4 Over-collection of specimens by reptile and frog keepers, and hunters within Australia.

In order of importance, the main conservation needs of reptiles and frogs in Australia are:

1 Habitat protection.

2 Elimination of 'pest' species.

3 Captive keeping, breeding and research.

4 Useful protective legislation for species and habitats.

5 Stopping smuggling and corruption within wildlife authorities.

Habitat Protection

In virtually all cases, reptile and frog species will not be threatened from collecting activities, shooting, etc., which could effectively be called 'harvesting a resource'. Collectors only remove about 30,000 reptiles from the wild annually for pets (and their food) on a national basis. Humans directly account for a further estimated 10 million herptile mortalities, mainly by running them over on the roads. Despite these numbers killed, the vast majority of specimens die without direct human influence, and are only threatened (on a species level) by permanent displacement through loss of habitat.

Populations appear to be able to withstand removal of specimens so long as enough specimens are left to support a viable breeding population. The cryptic nature of most species would make it virtually impossible to remove all specimens from most populations even if one consciously tried to do so.

Most reptiles and frogs do, however, have very specific habitat requirements. Many species can only survive in virgin habitats, including (for example) Red-crowned Toadlet *Pseudophryne australis*, Death Adders *Acanthophis* spp., Broad-headed Snake *Hoplocephalus bungaroides* and Rough-scaled Snake *Tropidechis carinatus*. Other species can only tolerate a certain amount of habitat modification before being eliminated. It should be mentioned here that a few species actually seem to benefit from certain types of habitat modification; usually from clearing vegetation or by providing a water supply. Some notable examples include Green Tree Frog *Litoria caerulea*, Grass Skink *Lampropholis guichenoti* and Common Brown Snake *Pseudonaja textilis*.

Most reptiles and frogs currently threatened are those species which have relatively strict habitat requirements, and whose habitats are under destruction. Examples include the Dwarf Form Copperhead *Austrelaps labialis* and Broad-headed Snake *Hoplocephalus bungaroides*.

Some 96 per cent of habitat within Australia has been modified since white settlement in 1788, and about 92 per cent of the land area is currently used for grazing or other farming activities. Currently not enough viable samples of various types of natural habitat are being preserved in the form of national parks, forestry areas and such like.

Although rainforests and some other threatened habitats are justifiably receiving attention and being protected (usually in insufficient amounts), other important habitats are simply being destroyed without even a whimper from most of the Australian public. Overgrazing and deliberate repeated burning is destroying most spinifex habitat and associated wildlife communities throughout many parts of northern inland Australia. Virtually no untouched bushland now exists in the drier half of New South Wales and Victoria, and many species of reptile, frog and other wildlife have all but disappeared from these areas. A number of other habitats are also being ruined without a rational assessment of the consequences.

Obviously 'progress' is necessary but Australia has an immense land area with relatively few people, and it should be relatively easy to preserve more areas in the form of national parks and similar. These in themselves are a recreational, scientific and cultural resource.

In many parts of Australia, the old saying 'If it moves, shoot it. If it doesn't, then chop it down' still applies. As the majority of Australia's area is used for farming practices it is important to educate farmers into realising the need to minimise excessive habitat modification, and bad farming practices, such as excessive clearing and overgrazing, which can lead to salinity, erosion and other problems. Those practices can contribute to a decline in reptile numbers over large areas, and it is in this area that more conservation resources should also be directed.

Elimination of Pest Species

Pest species are those which were introduced into the country by white settlers from overseas, and which have multiplied to become a threat to other wildlife and/or human agriculture.

Certain pest species eliminate reptiles and frogs and are a major threat to many species. These pests are particularly destructive as they can freely move into otherwise untouched areas and eliminate species from these places. For example, Water Buffaloes have trampled swamps in Kakadu National Park in the Northern Territory and elsewhere, eliminating many species of frog that would otherwise breed in the vegetation bordering these swamps. Even Crocodiles suffer by having their nests trampled by the Buffaloes.

Several other pests are having a major effect on herptile populations and may help to eliminate species. The Cane Toad *Bufo marinus* is eliminating frogs and frog-eating reptiles from most parts of Queensland, and nearby areas. The Mosquito Fish *Gambusia* spp. eliminates frogs by feeding on all types of tadpole. It is currently found near most populated parts of the country. Cats and foxes kill all species of wildlife, but reptiles are usually the dominant food, particularly in arid areas. They are found Australia wide. Rabbits, goats and stray farm animals remove ground vegetation and leave many species open to attack by birds and other predators. They are found in most of Australia. Pigs feed on eggs from nests of reptiles, particularly tortoises, turtles and crocodiles. They are found Australia wide. An enormous number of introduced plants displace native vegetation and change habitat sufficiently to cause certain species to die out. Some plant pests include Lantana, Privet, South African Boxthorn, Mimosa, Prickly Pear and Water Hyacinth.

These pest species should be eliminated. Despite a number of concerted campaigns by the author to get the federal government to take steps to exterminate the Cane Toad, before it wipes out more species, no action has yet been taken. Although (fortunately) no new pest species are being introduced into this country deliberately, as happened in the past, conservation efforts should be directed at biological means of removing current pests.

The Australian conservation movement seems to be largely blind to the threat posed by pest species to all forms of wildlife, although perhaps this 'blindness' is diminishing.

Captive Herptiles — Their Breeding and Research

In order to conserve reptiles and frogs, one needs to know about what one is trying to conserve. It is believed that some species of Australian herptile were made extinct even before they were discovered. A number of species are in serious decline for no apparent reason (e.g. frogs of the genus *Taudactylus*). Urgent research is needed on these species.

The more that is known about reptiles and frogs, the easier it becomes to develop useful conservation strategies based on sound scientific facts rather than emotions. Limited resources can be directed at conserving those species that are most endangered. Also with greater knowledge about our native herptiles it will become easier to take measures that will prevent more species from declining in numbers.

Insufficient government and other funds are at this stage allocated to herpetological research, and the training of professional herpetological workers, probably due to the relative lack of economic importance of most herptiles. More importantly, however, most research on reptiles and frogs has traditionally been done by unpaid amateur herpetologists and keepers. The problem in Australia is that activities by corrupt and misdirected fauna authorities have almost wiped out the amateur herpetological community. In 1973 there were an estimated 4000 amateur herpetologists and keepers within Australia. By 1980 the number was down to less than a thousand. In 1980–81 unconservationist influences within a Sydney herpetological society, sponsored by local wildlife officials, actually resulted in the disbandment of a previously important herpetological society. By the mid 1980s all Australian herpetological societies could only boast a combined membership of about 300 with memberships continuing to decline. Not only has research suffered as a result, but so too have other aspects of herptile conservation.

I believe that it is desirable for conservation reasons for as many people as possible to keep herptiles in captivity, either for research purposes or as pets. The primary reason is that by keeping herptiles people usually go through a logical progression that will only aid the conservation cause. First they learn more about herptiles. Then this knowledge is passed on to friends and others. As the public becomes more knowledgeable about herptiles they will understand the reasons to conserve them and, more importantly, will hopefully change their actions in relation to herptiles — too many Australians still kill snakes on sight. Some keepers will go on to do captive and/or field research and by publishing their findings more will come to be known about herptiles, further aiding conservation planning. Almost all professional herpetologists made the decision to embark on a herpetological career *after* keeping herptiles 'as pets'.

Many species of herptile *will* become extinct in the wild as a result of irreparable habitat destruction or for some other reason. Remaining specimens in the wild that are obviously going to become extinct should be caught and captive-bred and, if appropriate, relocated elsewhere. For some species the only way they are going to survive is by captive breeding. (The Round Island Boas are a classic case, where fast actions by herpetologists in the Northern Hemisphere have saved two species by captive breeding. The Boas had had their entire habitat destroyed by feral goats and introduced rodents.) All species of reptile *can* be bred in captivity.

Captive breeding can also serve to repopulate areas where a species has been exterminated or its population has dropped below natural replacement level. Extreme care should be taken when considering releasing any herptile into the wild. Ideally the herptile should only be released where it or its parents originated. If this is not possible then a specimen should only be released in an area containing genetically similar specimens and where the specimen will survive. A specimen should never be released too far from its point of origin as this may affect the local gene pool and play havoc with scientific records should it or its offspring be recaptured. If for some reason it is not possible to release a specimen close to its point of origin, one should attempt to find a captive home for the specimen or perhaps kill it. State museums always are in need of specimens. If the species is rare or endangered, it goes without saying that no specimen should ever be killed.

Embarking on a successful large-scale captive breeding programme to save a given species usually involves at some stage several hundred specimens. Usually no one person or organisation has the resources nor inclination to hold such a large number of specimens, so it becomes important to distribute the specimens around a large pool of keepers. Australia currently lacks such a pool, and I believe that developing one should be a major conservation objective.

Most research on herptile biology, including feeding and breeding behaviour, can *only* be done on captive specimens. I and a few others have held viable breeding colonies of a few types of herptile, usually snakes; however, most captive-breeding programmes relating to rare and endangered herptiles are being carried out in the United States and Europe. This includes Australian forms.

Assuming that rare or endangered species are not taken from the wild, taking specimens from the wild in a responsible manner for any reasonable purpose should be encouraged. Over-collecting of specimens by Australian herpetologists is sometimes a problem. However, when it occurs it usually only results in local extinctions of relatively widespread species. In the late 1960s much over-collecting of Broad-headed Snakes *Hoplocephalus bungaroides* occurred in areas near Sydney. However, even in this case the snakes would have probably been largely exterminated within ten years through habitat destruction — the removal of bush rock.

Those who claim that keeping herptiles in captivity is always cruel are wrong. Properly kept herptiles are infinitely healthier (and often have better sex lives) than their counterparts in the wild. After all, properly kept captive herptiles have plenty of food, no predators, no internal parasites and live in optimal temperatures.

Protective Legislation

When drafting herptile protection legislation, all Australian states have tried to keep in line with one another, and consequently all have similar rules and regulations. Although there is a need to protect reptiles, current legislation in force in all states is not appropriate for several reasons.

Current legislation relies on two main points. These are:

1 A prohibition on catching and keeping all or most species.

2 The state fauna authorities issuing permits to take, keep or kill herptiles. This entails herpetologists' finest details going on to a central register.

Most species of herptile in Australia are not endangered and usually common where they occur. With the exception of a few species, none are of major economic value as a resource. Whether or not they have statutory protection has no significance to them or their conservation status. Issuing a blanket prohibition on these herptiles is strongly counterproductive. Most importantly, the huge number of public servants required to administer the 'prohibition' is a gross misdirection of resources which should be directed at more immediate conservation problems. With 'prohibition' most people are discouraged from keeping herptiles thereby reducing the long-term conservation effort. Those who still want to keep or do research on these species will have to direct valuable time and effort into going through bureaucratic red tape, which would be better spent doing research. Those who don't go through the correct channels may find themselves breaking laws and labelled as criminals for relatively innocuous activities. All of which does little for the conservation cause. Tactics of fauna officers in most states regularly put keepers, researchers, etc. outside the law even when complying with all regulations and directions given.

The species that are truly endangered do not receive the necessary attention when a blanket prohibition is in force. In the long term the situation becomes worse as the number of herpetologists reduces, and even less is known about the endangered species.

The permit issuing system, especially in states with a prohibition on the collecting and keeping of reptiles, puts all enthusiasts on to a single register administered by very few people. Because there is big money in reptile smuggling, and reptile conservation and keeping are not often in the public spotlight, the permit system as it stands is wide open to corruption, leading to huge problems.

Today corruption in the upper levels of State Fauna Authorities has been repeatedly alleged in at least three states (New South Wales, Queensland and South Australia), so it must be assumed to be a major potential problem.

In 1981, the Author took actions against officers of the New South Wales National Parks and Wildlife Service (NPWS), for the theft from his house of some fourteen snakes, files and other materials, on May 8th that year. An 'out of court' settlement arranged between the then NPWS director, Don Johnstone and my father resulted in some of the snakes being returned. Upon NPWS direction by Johnstone and Clive Jones also of NPWS, on July 31st, Terry Boyland, then a reptile keeper at Sydney's Taronga Zoo, returned to the author a bag containing a number of snakes. One of the snakes in the bag, a Spotted Python *Bothrochilus maculosus* from the Glasshouse Mountains, Queensland, had not been taken from the author's house, and was in fact a substitute for a male Ant-hill Python *Bothrochilus perthensis*.

The substitute Spotted Python had been seized from a collection at Faulconbridge, New South Wales, by NPWS officers, some weeks prior. The Farmer family, had been assured by the NPWS officials that all snakes taken from them would in fact be released in the wild. The whole case was documented in detail by Fia Cumming, on August 25th, in *The Australian*, pages 1 and 2.

The key to new legislation should be to 'de-protect' all but the endangered species. Immediately most reptile and frog keepers would not have all their details in a central register and they would become considerably less vulnerable to the predations of overzealous fauna officers, break-ins by smugglers, etc. 'De-protection' would enable money to be spent elsewhere on conservation, and simultaneously encourage new people to enter the field of herpetology, further aiding conservation.

Although permits may still be required for endangered species, relatively few people would be covered by these permits.

Currently in present state laws, all protected herptiles and captive-bred offspring remain effectively crown property. In order to increase the incentive to captive-breed rare species it should be legislated that the state have no control over captive-bred young. Breeders should have the right to buy and sell specimens that are captive-bred. A way would have to be found to prevent the problem of people obtaining stock from the wild and then claiming that it is 'captive-bred', at least for endangered species.

Current fauna legislation gives fauna officers considerable rights of entry to property. Under normal circumstances a law-abiding citizen would not object to fauna officers having the right of entry to check on fauna held. The following documented case (see references) is given as just one reason why laws should be amended to prevent fauna officers from entering a herptile keeper's house under any circumstances without taking the herptile keeper to court to get court approval to gain entry, for 'good reason'.

On 27 March 1984 some New South Wales fauna officers did a 'routine check' of my premises and snakes held (all legally held, of course). I didn't realise at the time that my house was being 'cased' for a break-in. On 10 July 1984 the same men broke into the house, smashing doors, etc., and stole snakes, files, cash, computer disks and camera equipment. The fauna officers, who initially denied breaking in, were only caught afterwards because neighbours recognised them from the previous 'inspection'. The break-in was also reported on national television.

Perhaps more importantly some form of habitat protection legislation should be enacted to protect the habitats of endangered herptiles, as ultimately this may be the key to the survival of many given species. It is incredible that even now some species are being knowingly wiped out due to destruction of their last remaining habitat, often by government instrumentalities which are immune from prosecution. A classic example is the Loveridge's Frog *Philoria loveridgei*, of northern New South Wales, which is currently threatened by logging of its habitat. Unfortunately the drafting of any 'corruption-resistant' habitat protection legislation will be a very difficult task.

The alleged activities of officers of Australian fauna authorities that are not consistent with accepted conservation practices, and the mechanics of smuggling wildlife out of Australia, have been well-documented elsewhere and interested readers should seek further information in these areas.

Corruption and Smuggling

Specific cases of activities by Fauna authority officers, inconsistent with accepted conservation ideals, have been dealt with in detail by numerous people in various publications including, *Notes From NOAH*, *The Courier Mail*, and television programmes including *Willesee*, and *Animal Traffic*. Some of these cases are cited in the relevant part of the references section on pages 234–35. Elsewhere I have made very detailed proposals relating to smuggling and again it is hoped that interested readers seek further information on this subject.

Wildlife smuggling, though illegal, is a booming business. For example, many reptiles such as the Diamond Python *Morelia spilota* are worth up to $US10,000 (1987). Contrary to popular belief, there is much more money to be made out of smuggling reptiles than birds.

The practice of smuggling is extremely cruel in all its aspects. It can involve corruption in several countries.

Smugglers often source their reptiles from genuine breeders, by breaking into their facilities and stealing whatever is required. The reptiles are X-rayed to sterilise them invisibly, thereby preventing captive breeding by the purchaser and preserving exorbitant market prices. Smugglers do all they can to prevent captive breeding anywhere, and they benefit by reptiles becoming rarer both in the wild and in captivity.

Wildlife smuggling in its current form must be stopped. Banning smuggling, and increasing fines and penalties, does not work. It only serves to increase corruption and make everything associated with it worse. The only way to stop the illegal export of reptiles is to legalise it.

The Australian reptiles that command high prices overseas are not often rare or endangered species. They tend merely to be larger species such as pythons, monitors, large skinks, etc. Allowing moderate numbers of these to leave the country will not harm local populations in any way. However, it will stop all the unsavoury aspects of current smuggling rackets by undermining the whole business. Legalised export of reptiles would also indirectly strengthen conservation efforts of genuinely rare and endangered species, in a number of ways.

Assuming that most indigenous reptiles become 'de-protected', it would be harder for smugglers to get details of who has what anyway, hopefully making all stages of the export operation more trouble than it would be worth.

Importantly, the legal export of wildlife should have a few important regulations. Knowing that wildlife exporting is prone to corruption, legislation should be as corruption-resistant as possible. All known 'de-protected' species should be allowed to be exported. No individual or proxy should be permitted to export or import more than a certain number of specimens in order to prevent large-scale traders and more corruption and cruelty emerging.

A duty of around $A300 (at 1987 values) per specimen should be imposed. This would be important in protecting the welfare of specimens exported. When people have to pay something relatively substantial for a given reptile, it can be fairly safely assured that they will take proper care of it, which cannot be guaranteed if prices are cheap. Australia should not supply the 'disposable' reptile pet market by allowing totally unrestricted export of reptiles. Tortoises from North Africa and elsewhere were virtually wiped out in the wild when millions were sold in the USA and Europe for ridiculously cheap prices, which encouraged people to buy them but not take proper care of them, resulting in rapid death of specimens.

A tariff means of regulating exports has several advantages over a strict permit system, including less paperwork and red tape, and revenue raised for Australia that wouldn't otherwise be raised. It should be channelled into herpetological projects only (so as to prevent later governments trying to use reptile exports as an economic activity).

Corruption within national parks and wildlife authorities, real or perceived, is seen by many as a major problem. Without doubt current policies of Australian fauna authorities are assisting wildlife smuggling rackets, possibly inadvertently. By scrapping most current herptile 'protective' legislation and bringing in the changes suggested, problems of potential corruption will be reduced. The importance of having corruption-free and honest fauna authorities for reptile and frog protection cannot be overstated.

In order to further reduce risks of corruption within state wildlife authorities, they should be where possible broken into smaller, more effective units, or departments, as I have previously proposed (elsewhere).

REPTILE HABITATS

ALTHOUGH REPTILES AND frogs have areas of 'distribution' they are often restricted to a particular type or types of habitat. The long-term conservation of these species will largely depend on whether suitable viable habitats remain for them.

When discussing each individual habitat I give a brief overview of reptiles and frogs present, and not other species. However, it should be realised that these reptiles and frogs are only a part of a wider ecosystem which must be conserved. Not all the habitats shown are virgin. Most habitats within Australia have been altered by man to some extent for farming and other activities. Some reptile and frog species have actually benefited from this.

The selection of habitats covered here is by no means comprehensive, but does give some indication of the diversity of habitats within Australia. Although some people seek to define habitat zones on the grounds of temperatures, rainfall, soils, vegetation, etc., this is not always practical as most areas would have characteristics intermediate between given strict criteria. Reptile distributions within Australia do not in general tend to fit within a few specific geographical zones.

The photographs of habitats and descriptions are given in the following somewhat arbitrary order:

    North Australian arid
    Dry tropical
    Wet tropical
    Dry temperate
    Wet temperate sub-tropical
    Alpine/cool temperate

For each photograph I have given the nearest town or landmark location, which can be seen on any reasonable state map.

572

Northern Australian Arid

South of Halls Creek, Western Australia

Fig. 572

This is one of the richest reptile habitats in Australia. The spinifex *Triodia* spp. and rocky country provide shelter and habitat for an enormous number of reptile species. Diurnal reptiles likely to be encountered include various dragons (Agamidae) and skinks of the genus *Ctenotus*. On the rocky hills *Varanus Kingorum* is common.

At night when driving through this area Northern Death Adders *Acanthophis praelongus*, Spinifex Snake-lizards *Delma nasuta* and various pythons are likely to be encountered.

Devil's Marbles, Northern Territory

Fig. 573

Again, the rocks and spinifex in this area combine to provide habitat for a huge number of reptile species. By day *Ctenotus* skinks abound, as does a variety of agamids. Collared Whip Snakes *Demansia torquata*, King Brown Snakes *Pseudechis australis* and the Western Brown Snake *Pseudonaja nuchalis* may also be encountered.

The most common nocturnal reptiles include Stimson's Python *Bothrochilus stimsoni*, the Half-girdled Snake *Simoselaps semifasciatus*, various Geckoes and Pygopids.

Curtin Springs, Northern Territory

Fig. 574

A number of pigmy monitors are found in this red sand and spinifex country. They feed on various Geckoes including Smooth Knob-tailed Gecko *Nephrurus levis*, Centralian Knob-tailed Gecko *Nephrurus laevissimus* and *Ctenotus* skinks. Sand Goannas *Varanus gouldii* and various agamids are often seen active during the day.

Womas *Aspidites ramsayi* are usually found at night foraging.

Centralian Bluetongues *Tiliqua multifasciata* also occur here.

Uluru/Ayers Rock, Northern Territory

Fig. 575

Overburning of the spinifex plains around this area has resulted in a relative decline in local reptile numbers. Reasonably common species here include *Delma tincta*, Hooded Scalyfoot *Pygopus nigriceps*, Fat-tailed Diplodactylus *Diplodactylus conspicillatus*, Centralian Knob-tailed Gecko *Nephrurus laevissimus* and Woma *Aspidites ramsayi*, all likely to be seen at night. Various monitors, agamids and skinks may be found active during the day.

573

574

575
576

South Australian–Northern Territory Border

Fig. 576

Gross overgrazing by stock, rabbits and other animals has removed most ground cover from this area. Consequently few reptiles live here. Large monitors and agamids are the species most likely to be encountered.

The most common nocturnal reptile is probably the Fat-tailed Diplodactylus *Diplodactylus conspicillatus*.

MacDonnell Ranges, Northern Territory

Fig. 577

Among the many reptiles to be found in the rocky hills are Ridge-tailed Monitors *Varanus acanthurus*, Stimson's Pythons *Bothrochilus stimsoni*, Centralian Carpet Snakes *Morelia spilota bredli*, various geckoes, skinks and dragons.

In the associated spinifex, Desert Death Adders *Acanthophis pyrrhus* occur, although most specimens are caught when crossing roads at night.

Mount Isa, Queensland

Fig. 578

The habitat here contains various rock-dwelling monitors including the Perenty *Varanus giganteus* and Ridge-tailed Monitor *Varanus acanthurus*. Olive Pythons *Bothrochilus olivaceus* are also common here, particularly in rocky areas near watercourses, where they feed mainly on mammals and birds.

Bynoe's Geckoes *Heteronotia binoei* are probably the most common gecko species found here, although a number of different species may be found.

577

Barkly Tableland, near Camooweal, Queensland

Fig. 579

The seemingly inhospitable black-soil plains found here and in other parts of northern inland Queensland support a number of unique reptile species. These include Earless Dragons *Tympanocryptis cephalus* and some monitors, all of which usually occupy burrows during the heat of the day.

The most common snakes found here include Western Brown Snakes *Pseudonaja nuchalis*, King Brown Snakes *Pseudechis australis* and Moon Snakes *Furina ornata*.

East of Camooweal, Queensland

Fig. 580

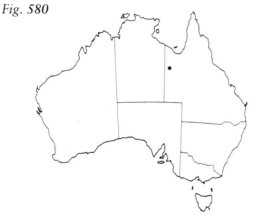

The spinifex *Triodia* grasslands and associated small termite mounds support a variety of reptiles. Common species include Centralian Bluetongue *Tiliqua multifasciata*, Stimson's Python *Bothrochilus stimsoni*, Spinifex-striped Gecko *Diplodactylus taeniatus*, Burton's Legless Lizard *Lialis burtonis* and Hooded Scalyfoot *Pygopus nigriceps*, all of which shelter in the *Triodia* clumps.

Agamids and various small skinks abound, and are usually seen running between the *Triodia* clumps.

Winton, Queensland

Fig. 581

The rocky mesa-type hills on the black-soil plains in inland Queensland support large numbers of reptiles. The species found on and adjacent to the hills are usually different from those found on the black-soil plains. Species found on the hills include Black-headed Python *Aspidites melanocephalus*, Perenty *Varanus giganteus* and various geckoes.

King Brown Snakes *Pseudechis australis* are common throughout this area.

579

580

581

Charleville, Queensland

Fig. 582

Overgrazing of this area has altered the species composition of reptiles found here. Species still likely to be found include Shingleback *Trachydosaurus rugosus* and Interior Bearded Dragon *Pogona vitticeps*, both of which are usually seen active or basking during the day, when the temperature is below 30°C. The Western Brown Snake *Pseudonaja nuchalis* is the most common snake here.

Dry Tropical

Ord River, Western Australia

Fig. 583

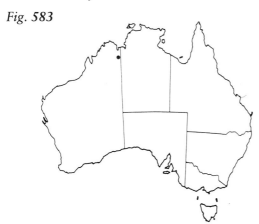

Reptiles are particularly abundant in the hilly rocky, *Triodia*-covered habitat here. This area has one of the highest numbers of individuals and some of the greatest species diversity in Australia.

In the hills by day agamids, monitors and skinks are seen actively foraging. At night the following species may be found: Children's Python *Bothrochilus childreni* (marked and unmarked specimens), Northern Death Adder *Acanthophis praelongus*, Half-girdled Snake *Simoselaps semifasciatus*, Hooded Scalyfoot *Pygopus nigriceps*, Burton's Legless Lizard *Lialis burtonis*, Brown Tree Snake *Boiga irregularis* and Splendid Tree Frog *Litoria splendida*, along with numerous geckoes and nocturnal skinks.

The river and adjoining flats support their own diverse assemblage of reptiles and frogs.

Turkey Creek, Western Australia

Fig. 584

The large termite mounds pictured here are found in scattered locations in the Pilbara, Kimberleys and elsewhere.

Within them a large number of reptiles take shelter and seek food. In this areas the reptiles found in termite mounds include Moon Snake *Furina ornata*, Children's Python *Bothrochilus childreni*, Black-headed Python *Aspidites melanocephalus*, Brown Tree Snake *Boiga irregularis*, King Brown Snake *Pseudeschis australis*, Fire-tailed Skink *Morethia ruficauda* and Broad-banded Sand Swimmer *Eremiascincus richardsoni*, along with various geckoes.

Victoria River, Northern Territory

Fig. 585

The open woodland pictured here supports a number of 'tropical' species. Dominant species include Frill-necked Lizard *Chlamydosaurus kingii*, Northern Bluetongue *Tiliqua intermedia*, Black-headed Python *Aspidites melanocephalus*, Children's Python *Bothrochilus childreni* and King Brown Snake *Pseudechis australis*.

The wet season is illustrated here, but at other times of the year the Bottle Trees loose their leaves, and most ground vegetation disappears.

582

583
584

585

586
587

Edith River Falls, Northern Territory

Fig. 586

The habitat within the escarpment country adjacent to the Edith River Falls is very similar in appearance to that of Sydney sandstone habitat, some 2000 km to the south-east. The herpetofauna of this area is in need of further investigation.

Reptiles found here include Bandy Bandy *Vermicella annulata*, Moon Snake *Furina ornata*, Black-headed Python *Aspidites melanocephalus*, and a wide variety of lizards.

Dominant frogs include Northern Holy Cross Frog *Notaden nichollsi* and Northern Water-holding Frog *Cyclorana australis*.

West Alligator River, Arnhem Highway, Northern Territory

Fig. 588

Along and within the river a number of semi-aquatic and aquatic reptiles and frogs will be found. The Saltwater Crocodile *Crocodylus porosus* makes it unsafe to swim in this and most other major northern rivers. Water Pythons *Bothrochilus fuscus* and Water Monitors *Varanus mertensi* prey on young crocodiles and other smaller animals.

A few different species of freshwater tortoise are also found here.

Katherine Gorge, Northern Territory

Fig. 587

The Katherine River supports a thriving population of Freshwater Crocodiles *Crocodylus johnstoni* and freshwater tortoises.

The adjacent hilly, rocky countryside supports a large number of reptiles. Some dominant species include Bynoe's Gecko *Heteronotia binoei*, Gilbert's Dragon *Lophognathus gilberti*, Black-headed Python *Aspidites melanocephalus*, Moon Snake *Furina ornata*, and King Brown Snake *Pseudechis australis*.

Arnhem Highway, Northern Territory

Fig. 589

The open woodland habitat of the Arnhem Highway supports a number of tropical and coastal species. The area is only lush and green during the wet season.

Species common here include Frill-necked Lizard *Chlamydosaurus kingii*, Spotted Tree Goanna *Varanus timorensis*, Children's Python *Bothrochilus childreni*, Carpet Snake *Morelia spilota variegata*, Brown Tree Snake *Boiga irregularis* and Green Tree Snake *Dendrelaphis punctulatus*. Most species, including the Frill-necked Lizard *Chlamydosaurus kingii*, appear to become more conspicuous during the wet season.

Fogg Dam, Northern Territory

Fig. 590

The swampy habitat here supports a huge number of frogs of various species including Northern Water-holding Frog *Cyclorana australis* and Marbled Frog *Limnodynastes convexiusculus*. Aquatic snakes, including Keelbacks *Amphiesma mairii* and Water Pythons *Bothrochilus fuscus*, are regularly seen moving amongst the partly submerged vegetation, water weeds, etc. Saltwater Crocodiles *Crocodylus porosus* are also found here.

North of Charters Towers, Queensland

Fig. 591

The rocky country around Charters Towers is where the Storr's Monitor *Varanus storri* is found. I have caught twenty specimens within an hour. The rocky habitat also supports Yellow-faced Whip Snakes *Demansia psammophis*, Spotted Pythons *Bothrochilus maculosus*, various agamids and geckoes.

Although Cane Toads *Bufo marinus* have decimated frog numbers, frog species found here include Northeastern Water-holding Frog *Cyclorana novaehollandiae* and Green Tree Frog *Litoria caerulea*.

West of Charters Towers, Queensland

Fig. 592

Open woodlands in inland northern Queensland support a number of monitors including Sand Goannas *Varanus gouldii*, and the Spotted Tree Goanna *Varanus timorensis*. Geckoes include Castelnaui's Gecko *Oedura castelnaui* and Bynoe's Gecko *Heteronotia binoei*.

Snakes found here include Black-headed Python *Aspidites melanocephalus*, Collared Whip Snake *Demansia torquata* and Curl Snake *Suta suta*.

588

589

590
591

592

593
594

Wet Tropical

Fitzroy Island, Queensland

Fig. 593

The islands of the Great Barrier Reef (and elsewhere) support their own herpetofauna. Lizards found on Fitzroy Island include Rainbow Skink *Carlia rhomboidalis*, Silver Skink *Sphenomorphus tenuis* and Eastern Snake-eyed Skink *Cryptoblepharus virgatus*.

The coral shallows of islands such as Fitzroy Island support large fish populations which in turn feed Sea Snakes and, to a lesser extent, Sea Turtles which inhabit the surrounding seas.

North of Townsville, Queensland

Fig. 594

The dry forest habitat north of Townsville has a wide variety of reptiles and frogs. In the lowlands Carpet Snake *Morelia spilota macropsila*, Water Python *Bothrochilus fuscus*, Keelbacks *Amphiesma mairii*, Green Tree Frog *Litoria caerulea*, Frilled-necked Lizard *Chlamydosaurus kingii* and Taipan *Oxyuranus scutellatus* are likely to be encountered.

On the virgin rocky hillsides Northern Death Adders *Acanthophis praelongus*, Zig-zag Geckoes *Oedura rhombifer*, Bynoe's Geckoes *Heteronotia binoei* and Spotted Pythons *Bothrochilus maculosus* occur.

Milla Milla, Queensland

Fig. 595

The Atherton Tableland, although located in tropical Queensland, has a more temperate climate because of its altitude. Watercourses, such as the one illustrated, provide breeding grounds for a number of frog species including Red-eyed Green Tree Frog *Litoria chloris*, Broad-palmed Frog *Litoria latopalmata* and a species related to Lesueur's Frog *Litoria lesueurii*. Eastern Water Dragons *Physignathus lesueurii* are common here.

Darker 'rainforest-phase' Carpet Snakes *Morelia spilota macropsila* occur in the adjacent forests and along watercourses in adjacent farming country.

595

Gordonvale, Queensland

Fig. 596

Coastal rainforests provide habitat for a huge number of frogs and reptiles. Frogs found in the forests shown include Green Tree Frog *Litoria caerulea*, Red-eyed Green Tree Frog *Litoria chloris*, Red-eyed Tree Frog *Litoria rothi*, Broad-palmed Frog *Litoria latopalmata* and Rocket Frog *Litoria nasuta*.

Scrub Pythons *Morelia amethistina*, Rough-scaled Snakes *Tropidechis carinatus*, Water Pythons *Bothrochilus fuscus* and Slatey-grey Snakes *Stegonotus cucullatus* also occur here.

Mon Repos Beach, Bundaberg, Queensland

Fig. 597

An important turtle-nesting rookery, this beach is also a major tourist attraction, relatively close to major population centres. At high tide during the night, Turtles move up the beach and then nest. The nests are dug and eggs laid in the grassy area above the high-tide line, between the trees and the beach.

Although Loggerheads *Caretta caretta* nest here in large numbers, Green Turtles *Chelonia mydas*, Flatbacks *Chelonia depressa* and Leathery Turtles *Dermochelys coriacia* are also known to nest here in smaller numbers.

In the adjacent seas, Sea Snakes are relatively common and easily found by divers.

Dry Temperate

Namoi River, Boggabri, New South Wales

Fig. 598

Four species of freshwater tortoise are known to occur here. These are Macquarie *Emydura macquarii*, Long-necked *Chelodina longicollis*, Broad-shelled *Chelodina expansa* and an undescribed *Elseya* species.

The adjacent river flats and river red gum (*Eucalyptus camaldulensis*) habitat supports Murray Valley form Carpet Snakes *Morelia spilota* subsp., Tree Skinks *Egernia striolata*, Tree Dtellas *Gehyra variegata* and Lace Monitors *Varanus varius*.

Between Boggabri and Manila, New South Wales

Fig. 599

The cleared grazing country of this area which was originally open woodland country still supports reptile and frog populations. Small skinks and geckoes constitute the most abundant herpetofauna. Species include Boulenger's Skink *Morethia boulengeri*, Eastern Bluetongue *Tiliqua scincoides* (inland New South Wales form), Bynoe's Gecko *Heteronotia binoei* and North-eastern Spiny-tailed Gecko *Diplodactylus williamsi*.

Snakes found here include Blue-bellied Black Snake *Pseudechis guttatus*, Eastern Brown Snake *Pseudonaja textilis* and occasionally Pale-headed Snake *Hoplocephalus bitorquatus*.

596

597
598

Gunnedah, New South Wales

Fig. 600

The rocky hills in the farming belt of western New South Wales have large numbers of reptiles. Geckoes include Barking Gecko *Underwoodisaurus milii*, Bynoe's Gecko *Heteronotia binoei* and Tree Dtella *Gehyra variegata*. Tree Skinks *Egernia striolata* often occur in large numbers. Snakes include Red-naped Snake *Furina diadema*, Bandy Bandy *Vermicella annulata*, Yellow-faced Whip Snake *Demansia psammophis* and Blind Snakes *Ramphotyphlops* spp.

Enngonia, New South Wales

Fig. 601

The desolate overgrazed plains near the New South Wales–Queensland border have only limited herpetofauna. The most regularly seen reptiles are Interior Bearded Dragon *Pogona vitticeps*, Shingleback *Trachydosaurus rugosus*, Western Brown Snake *Pseudonaja nuchalis* and King Brown Snake *Pseudechis australis*.

Small skinks of the genera *Ctenotus* and *Lerista* also occur here.

Bourke, New South Wales

Fig. 602

The mulga woodland in western New South Wales is home for many reptiles despite the fact that it is quite heavily grazed.

Geckoes found here include Eastern Spiny-tailed Gecko *Diplodactylus intermedius*, Tree Dtella *Gehyra variegata*, Tessellated Gecko *Diplodactylus tessellatus*, Steindachner's Gecko *Diplodactylus steindachneri*, Beaded Gecko *Lucasium daemium* and Bynoe's Gecko *Heteronotia binoei*.

Shinglebacks *Trachydosaurus rugosus* and Interior Bearded Dragons *Pogona vitticeps* are common.

Large elapids are the dominant snakes.

599

600

601

602

603

604
605

Wet Temperate/Sub-Tropical

Glasshouse Mountains, Queensland

Fig. 603

The rock outcrops and adjoining rainforests in combination harbour many species of reptile and frog from both rainforest and drier habitats.

The dominant gecko here is the Robust Velvet Gecko *Oedura robusta*. Spotted Pythons *Bothrochilus maculosus* are common on the rocky hillsides, while Carpet Snakes *Morelia spilota macropsila* are the most abundant snake elsewhere. Pink-tongued Skinks *Tiliqua gerrardii*, Small-eyed Snakes *Cryptophis nigrescens*, Stephen's Banded Snakes *Hoplocephalus stephensi* and Brown Tree Snakes *Boiga irregularis* are all likely to be found foraging at night.

Murwillumbah, New South Wales

Fig. 604

Where rainforest borders estuarine and brackish swamps many reptiles and frogs may be found. Nearby permanent and semi-permanent water is a breeding ground for the Common Froglet *Crinia signifera*, Northern Banjo Frog *Limnodynastes terraereginae* and the Brown-striped Frog *Limnodynastes peronii*.

Unfortunately in this area Cane Toads *Bufo marinus* have devastated populations of reptiles and frogs, yet Keelbacks *Amphiesma mairii* remain common because of their apparent immunity to the Cane Toad's poison.

Maroochydore, Queensland

Fig. 605

Lowland Paperbark swamp *Melaleuca* spp. provides excellent refuge for frogs and introduced Cane Toads *Bufo marinus*. Snakes found here include Keelbacks *Amphiesma mairii* and Carpet Snakes *Morelia spilota macropsila*. Red-bellied Black Snakes *Pseudechis porphyriacus* and Tiger Snakes *Notechis scutatus* have been completely eliminated from these habitats as Cane Toads invade.

Blacktown, New South Wales

Fig. 606

Cleared farming country typical of the Cumberland Plain near Sydney still provides habitat for a number of frogs and other reptiles. Dominant frogs include Brown-striped Frog *Limnodynastes peronii* and Common Froglet *Crinia signifera*. Long-necked Tortoises *Chelodina longicollis* may be found in water bodies.

The adjacent countryside supports Grass Skinks *Lampropholis guichenoti*, Eastern Bluetongue *Tiliqua scincoides* (coastal New South Wales form), Red-bellied Black Snake *Pseudechis porphyriacus* and Eastern Brown Snake *Pseudonaja textilis*.

606

607
608

West Head (near Salvation Creek), New South Wales

Fig. 607

St Ives, New South Wales

Fig. 608

This area typifies the sandstone country found to the north, south and west of Sydney and the rich herpetofauna within it. A number of unique species and forms live in the Sydney sandstone habitat.

Frogs found here include Giant Burrowing Frog *Helioporus australiacus*, Red-crowned Toadlet *Pseudophryne australis*, Green Tree Frog *Litoria caerulea*, Freycinet's Frog *Litoria freycineti* and Leaf Green Tree Frog *Litoria phyllochroa*.

Reptiles include White's Skink *Egernia whitii*, Cunningham's Skink *Egernia cunninghami* (sandstone form), Sand Goanna *Varanus gouldii*, Common Scalyfoot *Pygopus lepidopodus*, Diamond Python (Carpet Snake) *Morelia spilota spilota*, Death Adder *Acanthophis antarcticus*, Swamp Snake *Hemiaspis signata* and Golden-crowned Snake *Cacophis squamulosus*.

This area of 'typical Sydney bushland' was devastated by bushfire in 1980. Although most reptiles can tolerate some occasional bush burning, few species can tolerate excessively frequent burning of their habitat. Bushfires remove ground litter and protective cover for many reptiles, including Death Adders *Acanthophis antarcticus* and other snakes.

Some reptiles actually prefer freshly burnt habitat. In the habitat pictured, numbers of Red-throated Skinks *Leiolopisma platynotum* and Bearded Dragons *Pogona barbatus* seemed to increase after the burnoff as a result of specimens moving into the habitat from unburnt adjoining areas and from natural increase.

609

610

611

Alpine/Cool Temperate

Govett's Leap, Blackheath, New South Wales

Fig. 609

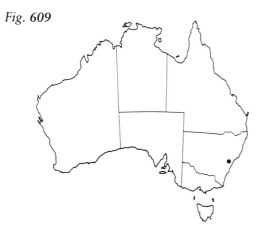

On top of the Blue Mountains escarpment pictured herpetofauna is a combination of alpine forms and Sydney sandstone forms. For example, three forms of Water Skink are known to occur in the area near Blackheath. These are Eastern Water Skink *Eulamprus quoyii*, Southern Water Skink *Eulamprus tympanum* (WTF) and Alpine Water Skink *Eulamprus kosciuskoi*.

In the Grosse River valley pictured here, frogs found include Barred Frogs *Mixophyes* spp., Lesueur's Frog *Litoria lesueurii*, Blue Mountains Tree Frog *Litoria citropa* and Leaf Green Tree Frog *Litoria phyllochroa*.

North of Lithgow, New South Wales

Fig. 610

The habitat here has a combination of sandstone, alpine and, to a minor extent, western slopes herpetofauna.

Lizards found here include Southern Water Skink *Eulamprus tympanum* (WTF), Black Rock Skink *Egernia saxatilis*, *Leiolopisma entrecasteauxii* (form A), Blotched Bluetongue *Tiliqua nigrolutea* (New South Wales alpine form) and Mountain Dragon *Amphibolorous diemensis*.

Snakes found here include Broad-headed Snake *Hoplocephalus bungaroides*, White-lipped Snake *Drysdalia coronoides* and Highland Copperhead *Austrelaps ramsayi*.

Most herptiles are caught when sheltering under slabs of sandstone.

Cox's River, Lithgow, New South Wales

Fig. 611

The hilly, open forest, granite country is typical of highland areas from Queensland through New South Wales to Victoria, and some similar country occurs in the far south-east of South Australia.

In the rock outcrops live Cunningham's Skinks *Egernia cunninghami* ('typical' form), often in large numbers, and these are the most commonly seen reptiles.

Other species found here include Common Bluetongue *Tiliqua scincoides* (western New South Wales form), Blotched Bluetongue *Tiliqua nigrolutea* (New South Wales alpine form), Three-fingered Burrower *Hemiergis decresiensis* and Hooded Snake *Unechis monachus*.

612

Bulls Head, Brindabella Ranges, Australian Capital Territory

Fig. 612

This forest habitat contains a typically alpine herpetofauna. A number of frogs are found here and the most

common reptiles are small skinks. These include Coventry's Skink *Leiolopisma coventryi*, *Leiolopisma entrecasteauxii* (form B), Spencer's Skink *Pseudemoia spenceri*, Maccoy's Skink *Hemiergis maccoyi*, Southern Water Skink *Eulamprus tympanum* (WTF) and three-lined Skink *Leiolopisma trilineata*.

Highland Copperheads *Austrelaps ramsayi* are the dominant snake.

Healesville, Victoria

Fig. 613

Although fewer reptiles than elsewhere are found in forest habitats in Victoria, those which do occur are specially adapted to the colder conditions usually present.

In the forest habitat pictured Lowland Copperheads *Austrelaps superbus* and White-lipped Snake *Drysdalia coronoides* are the dominant snake species. Other herptiles found here include Eastern Banjo Frog *Limnodynastes dumerilli*, Lesueur's Frog *Litoria lesueurii*, Blotched Bluetongue *Tiliqua nigrolutea* (southern form) and Garden Skink *Lampropholis delicata*.

REFERENCES

FURTHER INFORMATION ABOUT Australia's reptiles and frogs can be found from literally thousands of different sources. It is not practical to cite all references used in compiling this book, and a list of 'selected' references — those which most influenced the content of this text — has been given instead. (No references on the subject of photography are given.)

Important herpetological information can be published in 'popular' general works, in scientific literature (journals, etc.) and in the press. References have been grouped into several categories in the interests of simplicity, but it should be realised that the information given in some works might overlap these broad categories.

A relatively large number of references have been provided on the subjects of conservation, fauna authorities and smuggling, in order to support the conclusions drawn in this book. It is also a reflection of the importance of these subjects and of the fact that so little space has been devoted to them in this book.

Selected References — General Works

Banks, C. (1980), *Keeping Reptiles and Amphibians as Pets*, Thomas Nelson, Australia.

Barker, J. and Grigg, G. (1977), *A Field Guide to Australian Frogs*, Rigby, Australia.

Barrett, C. (1950), *Reptiles of Australia*, Cassell, Australia.

Bellairs, A. (1969), *The Life of Reptiles* (2 vols), Weidenfeld & Nicolson, London, UK.

Bustard, H.R. (1970), *Australian Lizards*, Collins, Australia.

Bustard, H.R. (1972), *Sea Turtles*, Collins, Australia.

Cann, J. (1978), *Tortoises of Australia*, Angus & Robertson, Australia.

Cann, J. (1986), *Snakes Alive! Snake Experts and Antidote Sellers of Australia*, Kangaroo Press, Australia.

Cogger, H.G. (1986), *Reptiles and Amphibians of Australia*, A.H. & A.W. Reed, Australia.

Clyne, D. (1969), *Australian Frogs*, Lansdowne Press, Australia.

Davey, K. (1970), *Australian Lizards*, Lansdowne Press, Periwinkle Series, Australia.

Frith, C. and D. (1987), *Australian Tropical Reptiles and Frogs*, Tropical Australian Graphics, Queensland, Australia.

Gibbons, W. (1983), *Their Blood Runs Cold: Adventures with Reptiles and Amphibians*, University of Alabama Press, Alabama, United States of America.

Goode, J. (1967), *Freshwater Tortoises of Australia and New Guinea*, Lansdowne Press, Australia.

Gow, G.F. (1977), *Snakes of the Darwin Area*, Museums and Art Galleries of the Northern Territory, Darwin, Australia.

Gow, G.F. (1982), *Australia's Dangerous Snakes*, Angus & Robertson, Australia.

Gow, G.F. (1983), *Snakes of Australia*, Angus & Robertson, Australia.

Griffiths, K. (1984), *Reptiles and Frogs of Australia*, View Productions, Australia.

Griffiths, K. (1987), *Reptiles of the Sydney Region*, Three Sisters Publications, Australia.

Guggisberg, C.A.W. (1972), *Crocodiles: Their Natural History, Folklore and Conservation*, Wren Publishing, Mount Eliza, Victoria.

Heatwole, H. (1976), *Reptile Ecology*, University of Queensland Press, St Lucia, Queensland.

Heatwole, H. (1987), *Sea Snakes*, New South Wales University Press in association with the Australian Institute of Biology, Australia.

Jenkins, R. and Bartell, R. (1980), *Reptiles of the Australian High Country*, Inkata Press, Australia.

Kinghorn, J.R. (1969), *The Snakes of Australia*, Angus & Robertson, Australia.

Krefft, G. (1869), *Snakes of Australia*, Australian Museum, Sydney, Australia.

Leutscher, A. (1976), *Keeping Reptiles and Amphibians*, David & Charles, UK.

Main, A.R. (1965), *Frogs of Southern Western Australia*, Handbook No.8, WA Naturalists Club, Perth, Australia.

McPhee, D.R. (1979), *The Observer's Book of Snakes and Lizards of Australia*, Methuen, Australia.

Mirtschin, P. and Davis, R. (1982), *Dangerous Snakes of Australia*, Rigby, Australia.

Noble, G.K. (1931), *The Biology of the Amphibia*, McGraw Hill, USA.

Schmida, G. (1985), *The Cold-blooded Australians*, Doubleday, Australia.

Schmidt, K.P. and Inger, R.F. (1957), *Living Reptiles of the World*, Hamish Hamilton, London, UK.

Stafford, P. (1986), *Pythons and Boas*, TFH Publications, Neptune, New Jersey, USA.

Storr, G.M. (1979), *Dangerous Snakes of Western Australia*, Western Australian Museum, Perth, Australia.

Sutherland, S.K. (1981), *Venomous Creatures of Australia*, Oxford University Press, Australia.

Swanson, S. (1976), *Lizards of Australia*, Angus & Robertson, Australia.

Tyler, M. (1976), *Frogs*, Collins, Australia.

Tyler, M. (1977), *The Frogs of South Australia*, South Australian Museum Publications, Adelaide, Australia.

Tyler, M. (1978), *Amphibians of South Australia*, Government Printer, South Australia.

Tyler, M., Smith, L.A., Johnstone, R.E. (1984), *Frogs of Western Australia*, Western Australian Museum, Perth, Australia.

Tyler, M. and Davies, M. (1986), *Frogs of the Northern Territory*, Government Printer, Northern Territory.

Waite, E.R. (1898), *Australian Snakes*, Thomas Shine, Sydney, Australia.

Waite, E.R. (1929), *The Reptiles and Amphibians of South Australua*, Government Printer, Adelaide, South Australia.

Worrell, E. (1969) *Dangerous Snakes of Australia and New Guinea*, Angus & Robertson, Australia.

Worrell, E. (1970) *Reptiles of Australia*, Angus & Robertson, Australia.

More Specific References — Journals, Etc

Frogs

Bailey, W.J. and Roberts, J.D. (1981), 'The bio-acoustics of the burrowing frog *Helioporus* (Leptodactylidea).' *Journal Natural History*, 15, pp. 693–702.

Hoser, R.T. (1982), 'Australian frog hunts.' *Journal NOAH*, 8 (1), pp. 27–42.

Humphries, R. (1979), 'Dynamics of a breeding frog community'. PhD thesis: Australian National University, Canberra.

Lee, A.K. (1967), 'Studies in Australian amphibia 2. Taxonomy, ecology and evolution of the genus *Helioporus* Gray (Anura: Leptodactylidae).' *Australian Journal of Zoology*, 15. pp. 367–439.

Littlejohn, M.J. (1963), 'Frogs of the Melbourne area.' *Victorian Naturalist*, 79 (10), pp. 296–304.

Martin, A.A. (1965), 'Tadpoles of the Melbourne area.' *Victorian Naturalist*, 82 (5), pp. 139–49.

Moore, J.A. (1961), 'The frogs of Eastern New South Wales.' *Bulletin American Museum of Natural History*, 121, pp. 149–386.

Tyler, M.J. (1974), 'First frog fossils from Australia.' *Nature*, 248 (5450), pp. 711–12.

Watson. G.F. and Littlejohn, M.J. (1985), 'Patterns of distribution, speciation and vicariance biogeography of south-eastern Australian amphibians', pp. 91–97 in *Biology of Australasian Frogs and Reptiles* (ed. G. Grigg, R. Shine and H. Ehman), Surrey Beatty & Sons in association with the Royal Zoological Society of New South Wales, Australia.

White, A. (1984), 'Zoogeography of Australian amphibians', pp. 283–89 in *Vertebrate Zoogeography and Evolution in Australia* (ed. M. Archer and G. Clayton), Hesperian Press, Sydney, Australia.

Reptiles (General)

Cogger, H.G. and Heatwole, H. (1981), 'The Australian reptiles: origins, biogeography, distribution patterns and island evolution', in (A. Keast, ed.) *Ecological Biogeography of Australia*, Dr W. Junk, Publisher, The Hague, Holland, pp. 1331–73.

Cogger, H.G., Cameron, E.E., and Cogger, H.M. (1983), *Zoological Catalogue of Australia*, Vol. 1, *Amphibia and Reptilia*. Bureau of Flora and Fauna, Canberra, Australia.

Gans, C. (ed.), (1969 *et.seq.*), *Biology of the Reptilia*, Vol. 1 (1969), Vol. 2 (1970), Vol. 3 (1970), Academic Press, London and New York.

Goin, C.J., Goin, A.B. and Zug, A.R. (1978), *Introduction to Herpetology*, W.H. Freeman & Co., San Francisco, USA.

Hoser, R.T. (1981), 'Reptiles of the Pilbara region (Western Australia).' *Journal of the Northern Ohio Association of Herpetologists (NOAH)*, 7 (1), pp. 12–32.

Hoser, R.T. (1982), 'Preferred activity temperatures of nocturnal reptiles in the Sydney Area.' *Herptile*, 9 (1), pp. 10–12.

Hoser, R.T. (1985), 'The truth about pelvic spurs on reptiles,' *Herptile*, 10 (3), p. 96.

Hoser, R.T. (1986), 'Reptile populations in the Gunnedah and Boggabri districts, NSW', *Journal of the Northern Ohio Association of Herpetologists (NOAH)*, 13 (1).

Hoser, R.T. (1987), 'More about pelvic spurs in Australian snakes and Pygopids', *Herptile*, 12 (2), pp. 65–67.

Houston, T.F. (1973), 'Reptiles of South Australia: a brief synopsis', in *South Australian Year Book, 1973*, Government Printer, South Australia.

Romer, A.S. (1956), *Oesteology of the Reptiles*, University of Chicago Press, Chicago, USA.

Storr, G.M. (1964), 'Some aspects of the geography of Australian reptiles', *Senkenbergiana Biologica*. Senkenbergische Naturforschende Gesellschaft, Frankfurt a.M., West Germany, 45, pp. 577–589.

Storr, G.M., and Harold, G. (1985), 'Herpetofauna of the Onslow region, Western Australia', *Records of the West Australian Museum*, 12 (3), pp 277–91.

Testudines

Georges, A. (1983), 'Reproduction in the Australian freshwater turtle *Emydura krefftii* (Chelonia: Chelidae)', *Journal Zoology*, London, 201, pp. 331–50.

Hoser, R.T. (1982), 'On breeding *Chelodina longicollis* (Shaw), with new cases of captive breeding and comments on more than one annual egg-laying season both in the wild and in captivity.' *Journal of the Northern Ohio Association of Herpetologists (NOAH)*, 8 (1), pp. 14–26.

Harless, M., and Morlock, H. (eds.) (1979), *Turtles: Perspectives and Research*, John Wiley & Sons, New York, USA.

Limpus, C.J. (1971), 'Sea turtle ocean-finding behaviour.' *Search*, 2, pp. 385–87.

Parmenter, C.J. (1980), 'The natural history of the Australian freshwater turtle *Chelodina longicollis* Shaw (Testudinata, Chelidae).' PhD thesis, University of New England, Armidale, NSW.

Pritchard, P.C.H. (1979), *Encyclopedia of Turtles*, TFH Publications, Neptune, New Jersey, USA.

Yntema, C.L. and Mrosovsky, N. (1980), 'Sexual differentiation in hatchling loggerheads (*Caretta caretta*) incubated at different controlled temperatures.' *Herpetologica*, 36, pp. 33–36.

Crocodiles

Bustard, H.R. (1971), 'Temperature and water tolerance of incubating crocodile eggs.' *British Journal of Herpetology*, 4, pp. 198–200.

Webb, G.J.W. (1977), 'The Natural History of *Crocodylus porosus*, 1. Habitat and Nesting' in *Australian Animals and Their Environment* (ed. Messel, H. and Butler, S.), Shakespeare Head Press, Sydney, Australia, pp. 239–84.

Webb, G.J.W., Messel, H. and Magnusson, W.E. (1977), 'The nesting biology of *Crocodylus porosus* in Arnhem Land, northern Australia.' *COPEIA* 1977, pp. 238–49.

Lizards

Badham, J.A. (1976), 'The *Amphibolorus barbatus* species-group (Lacertilia: Agamidae).' *Australian Journal of Zoology*, 24, pp. 423–43.

Bustard, H.R. (1968), 'The ecology of the Australian gecko *Heteronotia binoei* in northern New South Wales.' *Journal Zoology*, London, 156, pp. 483–97.

Copeland, S.J. (1945), 'Geographic variation in the lizard *Hemiergis decresiensis* (Fitzinger).' *Proceedings of the Linnaean Society of NSW*. 70, pp. 62–92.

Greer, A.E. (1974), 'The generic relationships of the Scincid Lizard Genus *Leiolopisma* and its relatives.' *Australian Journal of Zoology*, Supplement Series, 31, pp. 1–67.

Greer, A.E. (1979), 'A phylogenetic subdivision of Australian skinks.' *Records of the Australian Museum*, 32 (8), pp. 339–71.

Hoser, R.T. (1983), 'Notes on egg-laying in the Scalyfoot (*Pygopus lepidopodus*) and other reptiles.' *Herptile*, 8 (4), pp. 134–36.

Hoser, R.T. (1985), 'Notes on the feeding behaviour in the Common Scalyfoot (*Pygopus lepidopodus*), and the Burton's Legless Lizard (*Lialis burtonis*) in the Sydney district.' *Herptile*, 10 (3), pp. 93–94.

Kluge, A.G. (1974), 'A taxonomic revision of the lizard family Pygopodidae.' *Miscellaneous Publications of the Museum of Zoology, University of Michigan*, 147, pp. 1–221.

Kluge, A.G. (1967), 'Higher taxonomic categories of Gekkonid Lizards and their evolution.' *Bulletin of the American Museum of Natural History*, 135 (1), pp. 1–59.

Kluge, A.G. (1967), 'Systematics, phylogeny and zoogeography of the lizard genus *Diplodactylus* Gray (Gekkonidae).' *Australian Journal of Zoology*, 15, pp. 1007–108.

Mitchell, F.J. (1950), 'The Scincid genera *Egernia* and *Tiliqua* (Lacertilia).' *Records of the South Australian Museum*, 9 (3), pp. 275–308.

Moritz, C. (1983), 'Parthenogenesis in the endemic Australian lizard *Heteronotia binoei* (Gekkonidae).' *Science*, 222, pp. 735–37.

Pianka, E.R. (1972), 'Zoogeography and speciation of Australian desert lizards: An ecological perspective.' *COPEIA* 1972 (1), pp. 127–45.

Rawlinson, P.A. (1975), 'Two new lizard species from the genus *Leiolopisma* (Scincidae, Lygosominae), in South Eastern Australia and Tasmania.' *Memoirs of the National Museum, Victoria*, Melbourne, 36 (2), pp. 1–16.

Storr, G.M. (1963), 'The Gekkonid Genus *Nephrurus* in Western Australia, including a new species, and three new subspecies.' *Journal of the Royal Society of Western Australia*. 46 (3), pp. 85–90.

Storr, G.M. (1964), '*Ctenotus*, a new generic name for a group of Australian skinks.' *Western Australian Naturalist*, Perth, (9), pp. 84–88.

Storr, G.M. (1980), 'The monitor lizards (genus *Varanus* Merrem, 1820) of Western Australia.' *Records of the Western Australian Museum*, 8, pp. 237–93.

Storr, G.M., Smith, L.A. and Johnstone, R.E. (1981), *Lizards of Western Australia, 1. Skinks*. University of Western Australia Press and Western Australian Museum.

Throckmorton, G.S. *et. al.* (1985), 'Mechanism of frill erection in the Bearded Dragon *Amphibolorus barbatus*, with comments on the Jacky Lizard A. *muricatus* (Agamidae).' *Journal Morphology*, 183, pp. 285–92.

Snakes

Campbell, C.H. (1976), *Snake Bite, Snake Venoms and Venomous Snakes of Australia and New Guinea. An Annotated Bibliography*, Commonwealth Department of Health, School of Public Health and Tropical Medicine, Canberra.

Cogger, H.G. (1971), 'The venomous snakes of Australia and Melanesia' in *Venomous Animals and Their Venoms* (Bucherl, W. and Buckley, E., eds), vol 2, chapter 23. Academic Press, New York.

Coventry, A.J. and Rawlinson, P.A. (1980), 'Taxonomic revision of the elapid snake genus *Drysdalia* Worrell, 1961.' *Memoirs of the National Museum Victoria*. 41, pp. 65–78.

De Haas, C.P.J. (1950), 'Checklist of the snakes of the Indo-Australian archipelago.' *TREUBIA*. 20 (3), pp. 511–625.

Gillam, M.W. (1979), *The genus* Pseudonaja *(Serpentes: Elapidae) in the Northern Territory*. Territory Parks and Wildlife Commission, Research Bulletin No. 1.

Gow, G.F. (1981), 'A new species of python from central Australia.' *Australian Journal of Herpetology*, 1, pp. 29–34.

Hoser, R.T. (1980), 'Further records of aggregations of various species of Australian snake.' *Herpetofauna*, 12 (1), pp. 16–22.

Hoser, R.T. (1981), 'Unsuitable food item taken by a Death Adder *Acanthophis antarcticus* (Shaw).' *Herpetofauna*, 13 (1), pp. 30–31.

Hoser, R.T. (1981), 'Australian Pythons, part 1, Genera *Chondropython* and *Aspidites*.' *Herptile*, 6 (2), pp. 10–16.

Hoser, R.T. (1981) 'Australian Pythons, part 2, the smaller *Liasis*.' *Herptile*, 6 (3), pp. 13–19.

Hoser, R.T. (1981) 'Australian Pythons, part 3, the larger *Liasis*.' *Herptile*, 6 (4), pp. 3–12.

Hoser, R.T. (1982) 'Australian pythons, part 4, genus *Morelia* and *Python carinatus*, followed by discussions on the taxonomy and evolution of Australasian pythons.' *Herptile*, 7 (2), pp. 2–17.

Hoser, R.T. (1982), 'Mating behaviour of Australian Death Adders, Genus *Acanthophis* (Serpentes: Elapidae).' *Herptile* 7 (3), pp. 20–26.

Hoser, R.T. (1984) 'Search for the Death Adder.' *Notes from NOAH*. 11 (9), pp. 12–14.

Hoser, R.T. (1985), 'Genetic composition of Death Adders (*Acanthophis antarcticus*) (Serpentes: Elapidae) in the West Head area.' *Herptile*, 10 (3), p. 96.

Hoser, R.T. (1985), 'On melanistic tendencies in Death Adders *Acanthophis antarcticus* (Shaw).' *Litteratura Serpentium*, 5 (4), pp. 157–59.

Hoser, R.T. (1985), 'On the question of immunity of snakes.' *Litteratura Serpentium*. 5 (6), pp. 219–32.

Hoser, R.T. (1988), 'Some further comments in relation to Australian Pythons (Boidae) and their classification.' *Herptile* 13 (1), pp. 13–18.

Hunt, R.A. (1947), 'A key to the identification of Australian snakes.' *Victorian Naturalist*. 64, pp. 1–9.

Longmore, R. (ed.) (1986), *Atlas of Elapid Snakes of Australia*. Australian Government Publishing Service, Canberra.

McDowell, S.B. (1969), 'Notes on the Australian Sea-Snake *Ephalophis greyi* M. Smith (Serpentes: Elapidae: Hydrophiinae) and the origin and classification of Sea Snakes.' *Zoological Journal of the Linnaean Society*. 48, pp. 333–49.

McDowell, S.B. (1975), 'A catalogue of the snakes of New Guinea and the Solomons, with special reference to those in the Bernice P. Bishop Museum. Part 2. Anilioidea and Pythonidae.' *Journal of Herpetology*, 9 (1), pp. 1–79.

Mirtschin, P.J. (1983), 'Seasonal colour changes in the inland taipan *Oxyuranus microlepidotus.*' *Herpetofauna*, 14 (2), pp. 97–99.

Schwaner, T.D. (1985), *Snakes in South Australia: A Species List and Overview*, South Australian Museum, Adelaide, South Australia.

Seigal, R.A., Collins, J.T. and Novak, S.S. (1987), *Snakes: Ecology and Evolutionary Biology*, Macmillan Publishing Company, New York, USA.

Shine, R. (1985), 'Ecological evidence on the phylogeny of Australian Elapid snakes' in *Biology of Australasian Frogs and Reptiles* (eds Grigg, G., Shine, R. and Ehman, H.), Royal Zoological Society of NSW, pp. 255–60.

Shine, R. (1987), 'Ecological ramifications of prey size: Food habits and reproductive biology of Australian Copperhead Snakes (*Austrelaps*, Elapidae)', *Journal of Herpetology*, 21 (1), pp. 21–28.

Shine, R. (1987), 'Food habits and reproductive biology of Australian snakes of the Genus *Hemiaspis* (Elapidae)', *Journal of Herpetology*, 21 (1), pp. 71–74.

Smith, L.A. (1981), 'A revision of the Python genera *Aspidites* and Python (Serpentes: Boidae) in Western Australia', *Records of the Western Australian Museum*, 9, pp. 211–26.

Smith, L.A. (1985), 'A revision of the *Liasis childreni* species group (Serpentes: Boidae)', *Records of the West Australian Museum*, 12 (3), pp. 257–76.

Sutherland, S.K. (1981), *Venomous Creatures of Australia*, Oxford University Press, Australia.

Waite, E.R. (1918), 'Review of the Australian Blind Snakes (Family Typhlopidae)', *Records of the South Australian Museum*, 1, pp. 1–34.

Zulich, A.W. (1987), 'Juvenile colour in *Chondropython viridis.*' *Notes from NOAH*, 14 (9), pp. 16–17.

Hoser, R.T. (1984), 'A system for accounting for snakes', *Notes from NOAH*, 11 (7), pp. 10–14.

Hoser, R.T. (1987), 'Notes on the breeding of Death Adders *Acanthophis antarcticus*', *Herptile*, 12 (2), pp. 56–61.

Marcus, L.C. (1981), '*Veterinary Biology and Medicine of Captive Amphibians and Reptiles*,' Lea & Febiger, Philadelphia, USA.

Mirschin, P.J. (1985), 'An overview of captive breeding of Common Death Adders, *Acanthophis antarcticus* (Shaw) and its role in conservation' in *Biology of Australasian Frogs and Reptiles* (eds G. Grigg, R. Shine and H. Ehman), Royal Zoological Society of New South Wales.

Murphy, J.B. and Collins, J.T. (eds) (1980), *Reproductive Biology and Diseases of Captive Reptiles*. Society for the Study of Amphibians and Reptiles (SSAR).

Reichenbach-Klinke, H. and E. Flkan (1965), *Diseases of Reptiles*, TFH Publications, Neptune, New Jersey, USA.

Riches, R.J. (1976), *Breeding Snakes in Captivity*, Palmetto Publishing Company, St Petersburg, Florida, USA.

Ross, R.A. (1978), *The Python Breeding Manual*, Institute for Herpetological Research, California, USA.

Slavens, F.L. (1986), *Inventory of Live Reptiles and Amphibians in Captivity*', self published, Seattle, USA (annual publication).

Townson, S. and Lawrence, K. (eds) (1985), *Reptiles: Breeding, Behaviour and Veterinary Aspects*, British Herpetological Society.

Worden, A.N. and Lane Petter, W. (eds) (1957), *The UFAW Handbook on the Care and Management of Laboratory Animals*, Universities Federation for Animal Welfare, London.

Zimmerman, E. (1986), *Breeding Terrarium Animals*. TFH Publications, Neptune, New Jersey, USA.

Herptiles in Captivity

Barnett, B. (1980), 'Captive breeding and a novel egg incubation technique of the Children's Python (*Liasis childreni*)', *Herpetofauna*, 11 (2), pp. 15–18.

Bartlett, R.D. (1987), 'Some random thoughts on hobbyists', *Notes from HOAH*, 14 (6), 1987, pp. 3–6.

Cooper, J.E. and Jackson, O.F. (1981), *Diseases of the reptilia*, Academic Press, London.

Coote, J. (1978), 'Feeding captive snakes', *Herptile*, 3 (1), pp. 13–17.

Frye, F.L. (1981), *Biomedical and surgical aspects of captive reptile husbandry*, V.M. Publications, Edwardsville, Kansas, USA.

Hammond, S. (1988), 'A journey through the past and herpetoculture in the 'new age', *Notes from NOAH*, 15 (5), pp. 13–15.

Hay, M. (1971), 'Notes on the breeding and growth rate of *Morelia spilotes spilotes*', *Herpetofauna*, 3 (10).

Hay, R. (1987), 'An incubation technique for turtle eggs', *Notes from NOAH*, 14 (10), pp. 13–16.

Heijden, B.V.D. (1986), 'The husbandry and breeding of *Chondropython viridis*', *Litteratura Serpentium* (English edition), 6 (1), pp. 4–12.

Hoser, R.T. (1982), 'Frequency of sloughing in captive *Morelia, Liasis* and *Acanthophis* (Serpentes)', *Herptile*, 7 (3), pp. 20–26.

Conservation and Fauna Authorities

Anonymous (1986), 'Two on perjury charge over addict's death' *Sydney Morning Herald*, March 12, 1986.

Anonymous (1986), 'Customs officers charged' *Melbourne Age*, July 20, page 5.

Anonymous (1986), 'Zoo Knight 'was fauna trafficker'' *Daily Telegraph*, Sydney, Australia, May 7, page 13.

Anonymous (1987), 'Dead bird expert 'innocent'' *Times on Sunday*, February 8.

Bartlett, R.D. (1987), 'Some random thoughts on hobbyists', *Notes from NOAH*, 14 (6), pp. 3–6.

Belmore, B. (1981), 'Fish and Wildlife Scam Snares 25', *Pet Business*, 3 pp.

Begley, S. and Hager, M. (1981), 'The "Snakescam" Sting', *Newsweek*, 27 July.

Bloomer, T. (1982), 'More Snakescam', *Notes from NOAH*, 9 (6), pp. 5–8.

Bottom, B. (1985), *Connections: Crime Rackets and Networks of Influence Down Under*. Macmillan, Australia.

Bottom, B. (1987), *Connections 2: Crime Rackets and Networks of Influence in Australia*. Macmillan, Australia.

California Government (1982), 'Action no. 55812 — Lilley v. California Dept. of Fish and Game' (court and associated documents), dated 29 March 1982.

Cumming, F. (1981), "Snakies' feel bite of tough new stand' *The Australian*, pages 1 and 2. 25 August.

Dodd, C.K., jr (1986), 'Importation of live snakes and snake products into the United States', *Herpetological Review*, 17 (4), pp. 76–79.

Donnelly, M. (1984), 'Things are sliding in Lawson St, Redfern' *Daily Telegraph*, Sydney, Australia, 21st June.

Ehmann, H. and Cogger, H. (1985), 'Australia's endangered herpetofauna: A review of criteria and policies' in 'Biology of Australasian Frogs and Reptiles, ed. by Gordon Grigg, Richard Shine and Harry Ehmann, Royal Zoological Society of New South Wales. pp. 435–447.

Forbes, M. (1985), 'Criminals slip past 'corrupt' computer' *Sydney Morning Herald*, August 9, page 1.

Groombridge, B. (1982), *The IUCN Red Data Book, Amphibia–Reptilia* (part 1), International Union for the Conservation of Nature, London, UK.

Groombridge, B. (1983), *World Checklist of Threatened Amphibians and Reptiles*, Nature Conservancy Council, London, UK.

Haupt, R. (1973), 'US strikes at our 'Swiss connection'' *Sydney Morning Herald*, pages 1 and 3.

Hickie, D. (1985), *The Prince and the Premier*, Angus & Robertson, Australia.

Hoser, R. (1977), 'Transcript of phone conversation with Paul Ludowici (Re-NPWS)' Dated October.

Hoser, R. (1981), 'Transcript of Australian Herpetological Society Meeting' Dated September 25.

Hoser, R. (In Press), 'Smuggling snakes out of Australia . . . How the system works' *Litteratura serpentium*.

Hoser, R.T. (1987), '*Bufo marinus* menace', *Wildlife Australia*, 24 (1), p. 33.

International Union for the Conservation of Nature, (1973), 'Convention on International Trade in Endangered Species of Wild Fauna and Flora' (CITIES), Special Supplement to *IUCN Bulletin*, 4 (3), 12 pp.

Kennedy, M. (1981), 'Endangered species are big business', *National Parks Journal, Australia*, pp. 12–13.

Lilley, T. (1981), 'The sting: who really got stung', *Notes from NOAH*, 8 (2), pp. 1–3.

Livingstone, T. (1987), 'Scientist on fauna theft charge dies' *Brisbane Courier Mail*, February 3.

Livingstone, T. (1987), 'Parks service admits it took scientist's $10' *Brisbane Courier Mail*, February 10, page 2.

McShane, J. (1970), *Operation Doughnut: Report and Papers*, Federal Customs Department, Australia.

Messel, H. (1980), 'Rape of the north', *Habitat* (Australian Conservation Foundation), April, pp. 3–6.

Miller, M. *et. al.* (1982), 'Notes (re "Herpetological holocaust")', *Notes from NOAH*, 9 (5), pp. 1–9.

Miller, S. (1982), letter (re reptile smuggling), *Notes from NOAH*, 9 (4), pp. 7–9.

Mrosovsky, N. (1983), 'Conserving sea turtles', British Herpetological Society.

Myers, N. (1979), *Sinking Ark: A New Look at the Problem of Disappearing Species*, Pergamon Press.

Nichol, J. (1987), *The Animal Smugglers*, Christopher Helm (Publishers), London, UK.

NOAH (1981), 'ESHL gives a party . . . and the guests of honor don't show' and 'Australian National Parks and Wildlife outdoes US F and W', editorial, *Notes from NOAH*, 9 (1), pp. 1–7.

NOAH (1981), editorials and reprints (re corruption in fauna authorities), *Notes from NOAH*, 9 (3), pp. 1–13.

NOAH (1982), 'A different solution to "Snakecam"', editorial, *Notes from NOAH*, 9 (6), pp. 1–4.

NOAH (1987), 'Fisheries service proposes fatal experiments on endangered turtles', *Notes from NOAH*, 14 (12), pp. 16–17.

NOAH (1988), 'Herp smugglers get stung', editorial, *Notes from NOAH*, 15 (5), p. 1.

NSW Government (1982), 'NPWS v. Raymond Hoser' (court transcript and associated documents), dated 8 April 1982.

NSW Government (1984), 'NPWS v. Raymond Hoser' (court transcript and associated documents), dated 6 July 1984.

NSW Government (1984–85), 'NPWS v. Raymond Hoser' (court transcripts and associated documents), dated 25 July, 25 September, 7 December 1984, 27 March and 20 June 1985.

NSW Government (1985), 'Ackroyd v. NPWS' (court transcripts and associated documents), dated 26 June 1985.

NSW Government (1985), 'Hoser v. NPWS' (court transcripts and associated documents), dated September 1985.

Orders, R. and Bondy, A. (1988), 'Animal Traffic/Out of Australia' (Video Documentary), Cinecontact, UK.

Platt, C. (1974), 'Transport of live animals by air from Calcutta and Bangkok airports', report to International Society for the Protection of Animals.

Pritchard, P.C.H. (1986), 'In defense of private collections', *Herpetological Review*, 17 (3), pp. 56–58.

Purcell, R. (1978), 'Stolen Death Adder Unlikely to Bite' *North Shore Advocate Courier*, (Week unknown).

Rawlinson, P.A. (1980), 'Conservation of Australian amphibian and reptile communities' in *Proceedings of the Melbourne Herpetological Symposium* (ed. Banks, C.B. and Martin, A.A.), Zoological Board of Victoria, Melbourne, pp. 127–38.

Reddacliff, G.L. (1981), 'Necrotic enteritis in reptiles at Taronga Zoo', in *Proceedings of the Melbourne Herpetological Symposium* (ed. Banks, C.B. and Martin, A.A.), Zoological Board of Victoria, Melbourne. pp. 124–26.

Teese, P. (1985), 'Snakes at $600 a metre/Large scale smugglers slip through customs' *Sunday Telegraph*, Sydney, Australia, February 3, page 141.

Tyler, M.J. (ed.) (1979), *The Status of Endangered Australian Wildlife*, Royal Zoological Society of South Australia, Adelaide.

Whitton, E. (1986), *Can of Worms: A Citizen's Reference Book to Crime and the Administration of Justice*, Fairfax Library, Australia.

Willesee, M. (1984), 'Willesee' (News and Current Affairs), Dated July 10th 1984, (Video). Transmedia productions, Sydney, Australia.

Wilson, D., Murdoch, D. and Bottom, B. (1985), *Big Shots: A Who's Who in Australian Crime*, Sun Books, Australia.

Wilson, D., Robinson, P. and Bottom, B. (1987), *Big Shots 2*, Sun Books, Australia.

INDEX

Common names of species and scientific names are not cross-referenced in this index, but are cross-referenced throughout the text. For given species, only the page numbers of principal descriptions are given. Figure numbers of illustrations are given in the text at the appropriate places.

Acanthophis antarcticus 144–5
Acanthophis praelongus 145–6
Acanthophis pyrrhus 146–7
Acrochordidae 140
Acrochordus arafurae 140
Agamidae 56 ff
ailments 194–7
Alpine Water Skink 96
Amethystine Python: *see* Scrub Python
Amphibia 4–6
Amphibolurus diemensis 56–7
Amphibolurus muricatus 57
Amphiesma mairii 140–1
Anomalopus mackayi 83–4
Anomolopus swansoni 84
anorexia 194
Ant-hill Python 129
Arafura File Snake 140
Arnhem Highway, NT 215
Aspidites melanocephalus 123–4
Aspidites ramsayi 124–5
Austrelaps 147–9
Austrelaps labialis: see Austrelaps
Austrelaps ramsayi: see Austrelaps
Austrelaps superbus: see Austrelaps
Ayers Rock, NT 207

Bandy-bandy 174–5
Barking Gecko 79
Barkly Tableland, Qld 210
Barred Frog 29
Beaded Gecko 73
Bearded Dragon 64–5
Black-headed Monitor 120
Black-headed Python 123–4
Blackheath, NSW 228
Black Rock Skink 92
Black Tiger Snake 160–1
Blacktown, NSW 225
Black Whip Snake 152–3
Bleating Tree Frog 37
Blind Snakes 122–3
blisters 196
Blotched Bluetongue 109–10
Blue-bellied Black Snake 166–7
Blue Mountains Tree Frog 36–7
Boas 123

Bobtail: *see* Shingleback Lizard
Bog Eye: *see* Shingleback Lizard
Boggabri, NSW 220
Boidae 123 ff
Boiga irregularis 142
Bothrochilus albertisii 125–6
Bothrochilus childreni 126–7
Bothrochilus fuscus 127
Bothrochilus maculosus 127–8
Bothrochilus olivaceus 128–9
Bothrochilus perthensis 129
Bothrochilus stimsoni 129–30
Boulenger's Skink 105
Bourke, NSW 222
break-ins (thefts of reptiles) 205
breeding:
    frogs 6
    Tortoises and Turtles 8–9
Brindabella Ranges, ACT 229–30
Broad-banded Sand Swimmer 95
Broad-headed Snake 159
Broad-palmed Frog 38
Broad-shelled Tortoise 50–1
Broughton River Tiger Snake: *see* Black Tiger Snake
Brown-striped Frog 27–8
Brown Toadlet 31
Brown Tree Snake 142
Bufo marinus 22–3
Bufonidae 22
Bulls Head, ACT 229–30
Bundaberg, Qld 220
Bungarra: *see* Sand Goanna
Burnett's Skink 104
Burton's Legless Lizard 81
Bynoe's Gecko 72–3

Cacophis harriettae 149–50
Cacophis krefftii 150
Cacophis squamulosus 151
cages 181–2
Camooweal, Qld 210
Cane Toad 22–3
canker 195
captive breeding 189–93
captive herptiles (conservation) 203–4
capturing specimens 179
Caretta caretta 47
Carlia rhomboidalis 84–5

Carlia vivax 85
Carpentaria Whip Snake 173
Carpet Pythons 132–7
Centralian Bluetongue 109
Centralian Carpet Python: *see* Carpet Pythons
Centralian Knob-tailed Gecko 74–5
Central Netted Dragon 59
Challenger's Skink 99–100
Chappell Island Tiger Snake: *see* Black Tiger Snake
Charleville, Qld 212
Charters Towers, Qld 216
Chelidae 50 ff
Chelodina expansa 50–1
Chelodina longicollis 51–2
Chelodina oblonga 52
Chelodina rugosa 52–3
Chelonia depressa 47–8
Chelonia mydas 48–9
Cheloniidae 46–7
Children's Python 126–7
Chlamydosaurus kingii 58–9
Chondropython viridis 130–1
classification 3
Cogger's Velvet Gecko 75–6
Collared Whip Snake 153
Collett's Snake 165
Colubridae 140 ff
Colubrid snakes 140 ff
Common Eastern Blind Snake 122–3
Common Eastern Froglet 23
Common Scalyfoot 81
conservation 202–5
Copperheads 147–9
Copper-tailed Skink 88–9
corruption 205
Coventry's Skink 101
Cow Turd Skink: *see* Shingleback Lizard
Cox's River, NSW 229
Crinia signifera 23
Crocodilians 8
Crocodylus johnstoni 44
Crocodylus porosus 44
Crocodylidae 44 ff
Crowned Gecko 69
Cryptoblepharus plagiocephalus 85–6
Cryptoblepharus virgatus 86–7
Cryptodira 46–7

Cryptophis nigrescens 151–2
Ctenophorus decresii 59
Ctenophorus nuchalis 59
Ctenophorus pictus 60
Ctenotus leonhardii 87
Ctenotus pantherinus 87
Ctenotus regius 87
Ctenotus robustus 88
Ctenotus taeniolatus 88–9
Cunningham's Skink 89–90
Curl Snake 171
Curtin Springs, NT 207
Cyclorana australis 33
Cyclorana novaehollandiae 33–4

D'Albert's Python: *see* White-lipped Python
Death Adder 144–5
defence:
    lizards 11
    snakes 14–5
    Tortoises and Turtles 8
Delma molleri 79
Delma nasuta 80
Delma tincta 80–1
Demansia atra 152–3
Demansia psammophis 153
Demansia torquata 153
Dendrelaphis punctulatus 142–3
Denisonia devisi 153–4
Denisonia fasciata 154–5
Denisonia punctata 155
descriptions, introduction to 21
Desert Death Adder 146–7
Desert Tree Frog 42
Devil's Marbles, NT 207
De Vis' Banded Snake 153–4
Diamond Python: *see* Carpet Python
diet (tortoises and turtles) 9
Diplodactylus ciliaris 66
Diplodactylus conspicillatus 66–7
Diplodactylus intermedius 67–8
Diplodactylus steindachneri 69
Diplodactylus stenodactylus 69
Diplodactylus taeniatus 70
Diplodactylus tesselatus 70–1
Diplodactylus vittatus 71
Diplodactylus williamsi 71
Diporiphora superba 60–1

diseases 194–7
Dragon Lizards 56
Drysdalia coronoides 155
Drysdalia rhodogaster 155–6

Earless Dragon 66
Eastern Banjo Frog 25–6
Eastern Bluetongue 111–12
Eastern Brown Snake 169–71
Eastern Dwarf Tree Frog 37
Eastern Masters Snake 155–6
Eastern Short-necked Tortoise 56
Eastern Snake-eyed Skink 86–7
Eastern Snapping Tortoise 53–4
Eastern Spiny-tailed Gecko 67, 69
Eastern Tiger Snake 161–2
Eastern Water Dragon 63
Eastern Water Skink 96–7
Edith River Falls, NT 215
Egernia cunninghami 89–90
Egernia depressa 90
Egernia frerei 91
Egernia major 91
Egernia rugosa 91–2
Egernia saxatilis 92
Egernia stokesii 92–3
Egernia striolata 93
Egernia whitii 94
Elapidae 144 ff
elimination of pest species 203
Elseya dentata 53
Elseya latisternum 53
Emydura krefftii 54
Emydura macquarii 55
Emydura signata 56
Enngonia, NSW 222
Eremiascincus richardsoni 95
Eretmochelys imbricata 49
Estuarine Crocodile: *see* Saltwater Crocodile
Eulamprus heatwolei : *see Eulamprus tympanum*
Eulamprus kosciuskoi 96
Eulamprus quoyii 96
Eulamprus tenuis 97
Eulamprus tympanum 97

fallacies about reptiles 18–19
Fat-tailed Diplodactylus 66–7
fauna authorities 198, 202–5
feeding:
   in captivity 184–6
   lizards 11–2
   snakes 14
   Tortoises and Turtles 9
fighting (lizards) 12
File Snakes 140
finding reptiles and frogs 177–9
Fire-tailed Skink 105
Fitzroy Island, Qld 219
Flatback Turtle 47–8
Fletcher's Frog 24–5
Fogg Dam, NT 210
food supplies for captive specimens 186
Freckled Monitor 120

Freshwater Crocodile 44
Freshwater Snake: *see* Keelback
Freshwater Tortoises 50 ff
Freycinet's Frog 38
Frill-necked Lizard 58–9
frogs 4–6
Front-fanged Venomous Land Snakes 144
fungus 194
Furina diadema 156–7
Furina ornata 157

Garden Skink 100–1
Gastroenteritis 197
Geckoes 66 ff
Gehyra punctata 71–2
Gehyra variegata 72
Gekkonidae 66 ff
Giant Banjo Frog 26–7
Giant Burrowing Frog 23–4
Gidgee Skink 92–3
Gilbert's Dragon 61
Gippsland Water Dragon 62
Glasshouse Mountains, Qld 225
Goannas 113 ff
Golden-crowned Snake 151
Gonocephalus spinipes 61
Gordonvale, Qld 220
Gould's Monitor: *see* Sand Goanna
Govett's Leap, NSW 228
Grass Skink 100–1
Great Barred Frog 29–30
Green and Golden Bell Frog 35
Green Dragon 60
Green Python 130–1
Green Tree Frog 35
Green Tree Snake 142–3
Green Turtle 48–9
Gunnedah, NSW 222

habitat protection 202–3
habitats 206–30
Half-girdled Snake 171
Halls Creek, WA 206
handling reptiles 179–80
Hawksbill Turtle 49
Healesville, Vic 230
Helioporus australiacus 23–4
Hemiaspis signata 158
Hemiergis decresiensis 98
Hemiergis graciloides 99
Hemiergis maccoyi 99
Heteronotia binoei 72–3
Hooded Scalyfoot 83
Hooded Snake 173
Hoplocephalus bitorquatus 158–9
Hoplocephalus bungaroides 159
Hoplocephalus stephensi 160
hybrid pythons 137
Hydrophiidae 175–76
Hylidae 33 ff

infections, body 195
incubating reptile eggs 190–1
initiating breeding 190
Inland Taipan 162, 164
Interior Bearded Dragon 65–6

internal parasites 196–7
Island Tiger Snakes 160–1

Jacky 57

Katherine Gorge, NT 215
Keelback 140–1
keeping records 177
Keferstein's Frog 32
King Brown Snake 165
King Island Tiger Snake: *see* Black Tiger Snake
Krefft's Dwarf Snake 150
Krefft's Tortoise 54

Lace Monitor 120–1
Lampropholis challengeri 99
Lampropholis delicata 100
Lampropholis guichenoti 100–1
Lampropholis mustelina 101
Land Mullet 91
Lapemis hardwickii 175–6
Leaf-green Tree Frog 40
Lechriodus fletcheri 24
leeches 194
Legless Lizards 79
Leiolopisma coventryi 101
Leiolopisma entrecasteauxii 102
Leiolopisma platynotum 102–3
Leiolopisma trilineata 103
Leonhard's Skink 87
Lerista bougainvillii 103
Lerista muelleri 104
Lesueur's Frog 39
Lesueur's Gecko 76
Lialis burtonis 81
licences 198
Limnodynastes convexiusculus 25
Limnodynastes dumerilii 25–6
Limnodynastes dumerilii dumerilii: *see Limnodynastes dumerilii*
Limnodynastes dumerilii grayi: *see Limnodynastes dumerilii*
Limnodynastes interioris 26
Limnodynastes ornatus 27
Limnodynastes peronii 27–8
Limnodynastes tasmaniensis 28–9
Limnodynastes terraereginae 29
Lithgow, NSW 229
Litoria alboguttata 34–5
Litoria aurea 35
Litoria caerulea 35
Litoria chloris 36
Litoria citropa 36–7
Litoria dentata 37
Litoria fallax 37
Litoria freycineti 38
Litoria latopalmata 38
Litoria lesueuri 39
Litoria nasuta 39
Litoria peronii 39–40
Litoria phyllochroa 40–1
Litoria raniformis 41
Litoria rothi 41–2
Litoria rubella 42

Litoria splendida 42–3
Litoria verreauxii 43
Little Spotted Snake 155
Lizards 10–13, 56 ff
locomotion:
   frogs 5
   snakes 14
Loggerhead Turtle 47
Long-necked Tortoise 51–2
Lophognathus gilberti 61
Lucasium damaeum 73
Lygisaurus burnettii 104

Maccoy's Skink 99
Macdonnell Ranges, NT 209
Macquarie Tortoise 55
Mainland Taipan 164
Major Skink 91
Mangrove Monitor 116
Manila, NSW 220
Marbled Frog 25
Marine Toad: *see* Cane Toad
Maroochydore, Qld 222
Marsh Snake: *see* Swamp Snake
mating:
   lizards 12
   snakes 15–16
Mertens' Water Monitor 118
metamorphasing frogs 4
Milla Milla, Qld (habitat) 219
mites 195
Mixophyes balbus 29
Mixophyes fasciolatus 29–30
Monitor Lizards 133 ff
Mon Repos, Qld 220
Moon Snake 157
Morelia amethistina 131
Morelia carinata: *see Morelia spilota*
Morelia oenpelliensis 132
Morelia spilota 132–3
Morelia spilota bredli: *see Morelia spilota*
Morelia spilota imbricata: *see Morelia spilota*
Morelia spilota macropsila: *see Morelia spilota*
Morelia spilota spilota: *see Morelia spilota*
Morelia spilota sub. sp.: *see Morelia spilota*
Morelia spilota variegata: *see Morelia spilota*
Morethia boulengeri 105
Morethia ruficauda 105
Mountain Dragon 56–7
Mount Isa, Qld 209
mouth rot 195
Mueller's Skink 104
Murwillumbah, NSW 222
Myobatrachidae 23

National Parks and Wildlife Service: *see* fauna authorities
Namoi River, NSW 220
Neobatrachus sudelli 30
Nephrurus asper 74
Nephrurus laevissimus 74–5
Nephrurus levis 75

North-eastern Spiny-tailed Gecko 71
North-eastern Water-holding Frog 33
Northern Banjo Frog 29
Northern Bluetongue 108–9
Northern Death Adder 145–6
Northern Holy Cross Frog 30–1
Northern Long-necked Tortoise 52–3
Northern Snapping Tortoise 53
Northern Water-holding Frog 33–4
North-western Dwarf Bearded Dragon 65
Notaden nichollsi 30–1
Notechis ater 160–1
Notechis ater humphreysi: see *Notechis ater*
Notechis ater serventyi: see *Notechis ater*
Notechis ater sub. sp.: see *Notechis ater*
Notechis scutatus 161–2
NPWS: see fauna authorities

Oblong Tortoise 52
obtaining specimens 177
Oedura coggeri 75–6
Oedura lesueurii 76
Oedura rhombifer 76
Oedura robusta 77
Oedura tryoni 77
Oenpelli Python 132
Olive Python 128–9
Orange-naped Snake: see Moon Snake
Ord River, WA 212
origins of reptiles 1
Ornate Burrowing Frog 27
Oxyuranus microlepidotus 162, 164
Oxyuranus scutellatus 164

Painted Burrowing Frog 30
Painted Dragon 60
Pale-headed Snake 158–9
parasites 194
Pelamis platurus 176
pelvic spurs 189
Perenty 114–15
Peron's Tree Frog 39–40
pest species 203
photography 199–201
Phyllurus platurus 78
Physignathus howittii 62
Physignathus lesueurii 63–4
Pine Cone Lizard: see Shingleback Lizard
Pink-tongued Skink 107–8
Pleurodira 50 ff
Pogona barbatus 64
Pogona mitchelli 65
Pogona vitticeps 65–6
predation (tortoises and turtles) 8
preserving specimens 198
protective legislation 204–5
Pseudechis australis 165
Pseudechis colletti 165

Pseudechis guttatus 166
Pseudechis porphyriacus 167
Pseudemoia spenceri 106
Pseudonaja nuchalis 168–9
Pseudonaja textilis 169, 171
Pseudophryne australis 31
Pseudophryne bibronii 31
Pygmy Mulga Monitor 115
Pygmy Spiny-tailed Skink 90
Pygopodidae 79
Pygopus lepidopodus 81–2
Pygopus nigriceps 83
Python classification, problems 137
Pythoninae 123 ff
Pythons 123 ff

Racehorse Goanna: see Sand Goanna
Rainbow Skink 84–5
Ramphotyphlops nigrescens 122
Ramphotyphlops proximus 122–3
Red-bellied Black Snake 167
Red-crowned Toadlet 31
Red-eyed Green Tree Frog 36
Red-eyed Tree Frog 41–2
Red-naped Snake 156–7
Red-throated Skink 102–3
regulations 198, 202–5
reproduction (lizards) 12
Reptilia (class) 7–20
research 203–4
respiratory infections 195
Ridge-tailed Monitor 113–14
Robust Velvet Gecko 77
Rocket Frog 39
Rosen's Snake 154–5
Rough Knob-tailed Gecko 74
Rough-scaled Snake 172

Salientia 4–6, 22 ff
Saltwater Crocodile 44, 46
Sand Goanna 115–16
Sauria 11–13, 56 ff
scalation:
  lizards 13
  snakes 17–8
scale rot 195
Scincidae 83 ff
Scrub Python 131
Sea Snakes 175–6
Sea Turtles 46–7
Serpentes 13–20, 122
sexing 189
She-oak Skink 107
Shingleback Lizard 112–13
Siaphos equalis 106–7
Sideneck Tortoises 50
Silver Skink 97
Skinks 83
Slatey-grey Snake 143–4
Sleepy Lizard: see Shingleback Lizard
Small-eyed Snake 151–2
Smooth Knob-tailed Gecko 75
smuggling 205

snakebite 20
snakes 10, 13–20, 122 ff
Spine-bellied Sea Snake 175–6
Spotted Black Snake 166
Spotted Python 127–8
soft shell 195
South Australian–Northern Territory border 209
Southern Angle-headed Dragon 61
Southern Frogs 23
Southern Leaf-tailed Gecko 78
Southern Water Skink 97
specimens: see individual topics, e.g. capturing, photography etc.
Spencer's Skink 106
Spinifex Snake Lizard 80
Spinifex Striped Gecko 70
Spiny-tailed Gecko 66
Splendid Tree Frog 42–3
Spotted Black Snake: see Blue-bellied Black Snake
Spotted Dtella 71–2
Spotted Grass Frog 28–9
Spotted Tree Monitor 119
Squamata 10–20, 56, 122 ff
Stegonotus cucullatus 143
Steindachner's Gecko 69
Stephen's Banded Snake 160
Stimson's Python 129–30
St Ives, NSW 224
Stone Gecko 71
Storr's Monitor 119
Striped Burrowing Frog 34–5
Striped Skink 88
Swamp Snake 158

tail autotomy 11
Tawny Dragon 59
terrapins: see tortoises
Tesselated Gecko 70–1
Testudines 8–10, 46 ff
Thick-tailed Gecko: see Barking Gecko
Three-fingered Burrower 98
Three-lined Skink 103
Three-toed Skink 106–7
ticks 196
Tiliqua casuarinae 107
Tiliqua gerrardii 107–8
Tiliqua intermedia 108
Tiliqua multifasciata 109
Tiliqua nigrolutea 109–10
Tiliqua occipitalis 111
Tiliqua scincoides 111–12
toads 4, 22
tortoises 8–10, 46, 50 ff
Townsville, Qld 219
Trachydosaurus rugosus 112–13
transporting specimens 180
Tree Dtella 72
Tree Frogs 33
Tree Skink 93
Tropidechis carinatus 172
Tryon's Gecko 77
Turkey Creek, WA 212
turtles 8–10, 46 ff
Tympanocryptis cephalus 66

Typhlopidae 122

Uluru, NT 207
Underwoodisaurus milii 79
Unechis boschmai 173
Unechis monachus 173
Unechis spectabilis 174
Uperoliea laevigata 32
Uperoliea orientalis 32

Varanidae 113 ff
Varanus acanthurus 113–14
Varanus giganteus 114–15
Varanus gilleni 115
Varanus gouldii 115–16
Varanus gouldii flavirufus: see *Varanus gouldii*
Varanus gouldii gouldii: see *Varanus gouldii*
Varanus gouldii sub. sp.: see *Varanus gouldii*
Varanus indicus 116
Varanus kingorum 118
Varanus mertensi 118
Varanus storri 119
Varanus timorensis 119
Varanus tristis 120
Varanus tristis orientalis: see *Varanus tristis*
Varanus tristis tristis: see *Varanus tristis*
Varanus varius 120–1
Vermicella annulata 174
Victoria River, NT 212

Warty Green and Golden Bell Frog 41
Water Dragon: see Eastern Water Dragon
Water Monitor: see Mertens' Water Monitor
Water Python 127
Water Skink: see Eastern Water Skink
Weasel Skink 101
West Alligator River, NT 215
Western Bluetongue 111
Western Brown Snake 168–9
Western Snake-eyed Skink 85–6
Western Tiger Snake 160–1
West Head, NSW 224
Whistling Tree Frog 43
White-crowned Snake 149
White-lipped Python 125–6
White-lipped Snake 155
White's Tree Frog: see Green Tree Frog
White's Skink 94
Winton, Qld 210
Woma 124–5
Worm Snakes: see Blind Snakes

Yakka Skink 91–2
Yellow-bellied Sea Snake 176
Yellow-faced Whip Snake 153

Zig-zag Gecko 76